Youssef HAIRCH

Dynamics of Complex Fluids:

Youssef HAIRCH

Dynamics of Complex Fluids:

Approaches with Allen-Cahn and Navier-Stokes

ScienciaScripts

Imprint

Cover image: www.ingimage.com

This book is a translation from the original published under ISBN 978-620-6-71642-6.

Publisher:
Sciencia Scripts
is a trademark of
Dodo Books Indian Ocean Ltd. and OmniScriptum S.R.L publishing group

120 High Road, East Finchley, London, N2 9ED, United Kingdom
Str. Armeneasca 28/1, office 1, Chisinau MD-2012, Republic of Moldova, Europe
Printed at: see last page
ISBN: 978-620-7-91090-8

Summary

The coupling of the Navier-Stokes equations with the Allen-Cahn model represents a significant advance in the field of fluid mechanics and interface modeling. This development has its roots in the need to better understand and simulate complex multi-phase fluid dynamics. This book ***(Complex Fluid Dynamics: Approaches with Allen-Cahn and Navier-Stokes)*** *explores the approximation of dynamics between two fluids of different densities and viscosities. It presents a model that couples the Allen-Cahn equation, describing the evolution of a scalar-order parameter, with the time-dependent Navier-Stokes equations governing fluid motion. This coupling involves a surface tension term, which takes into account the energy required to create an interface between phases, proportional to the curvature of the interface, leading to the formation of a dynamic interface. The two-phase Navier-Stokes/Allen-Cahn equations simulate several fluid flows, using the Allen-Cahn model for diffuse interfaces and the Navier-Stokes equations for fluid dynamics.*

This work successfully simulates the generation of bubbles in a liquid subjected to an acoustic field, accurately predicting bubble size and formation frequency. Numerical results are in good agreement with experimental data.

BY THE WAY :

The study of phase separation in multiphase systems is of crucial importance to many fields of science and industry. Understanding the complex interactions between different fluid phases enables us to design innovative materials, improve manufacturing processes and solve critical environmental problems. However, the intrinsically complex nature of these phenomena requires advanced modeling tools and sophisticated numerical techniques to be fully understood and exploited.

Our thanks go to the many researchers and engineers whose pioneering work laid the foundations of this fascinating field. Their dedication to the advancement of knowledge and technology has made possible the development of the methods presented in this paper. We hope that this work will inspire further research and innovation in the field of computational fluid dynamics.

Table of contents

Foreword

The study of dynamics between two fluids of different densities and viscosities is a fascinating and complex field, with multiple applications ranging from industrial engineering to fundamental research in fluid physics. This book explores the subject in depth, presenting an innovative model that combines the Allen-Cahn equation and the time-dependent Navier-Stokes equations.

The Allen-Cahn equation, used to describe the evolution of a scalar-order parameter, is coupled here to the Navier-Stokes equations, which govern fluid motion under the influence of various forces. This coupling is based on a surface tension term, crucial for understanding the energy required to create an interface between phases. This term, proportional to the curvature of the interface, enables the formation of a dynamic interface. The two-phase Navier-Stokes/Allen-Cahn equations developed in this book offer an accurate simulation of multiple fluid flows. The Allen-Cahn model is used to represent diffuse interfaces, while the Navier-Stokes equations describe fluid dynamics. A notable application of this model is the simulation of bubble generation in a liquid subjected to an acoustic field. Predictions of bubble size and formation frequency are highly accurate and in good agreement with available experimental data. The results presented in this book are the fruit of rigorous and thorough research. We hope they will contribute to a better understanding of the complex phenomena linked to interactions between fluids of different natures, and inspire new investigations in this rapidly evolving field.

This book is aimed at researchers, engineers and students wishing to deepen their knowledge of fluid dynamics and associated mathematical models.

Enjoy your reading.

HAIRCH Youssef

Chapter I: Mathematical model

A little history

The need to simulate multi-phase fluid dynamics, where interfaces between different phases play a crucial role, has led to the integration of Allen-Cahn models with the Navier-Stokes equations. The coupling involves a surface tension term in the Navier-Stokes equations, which takes into account the energy required to create and maintain an interface between phases. This term is proportional to the curvature of the interface, making it possible to model complex, dynamic interfaces. The unsteady Navier-Stokes-Allen-Cahn (NS-AC) model combines the advantages of both approaches: the Navier-Stokes equations for fluid dynamics and the Allen-Cahn model for the representation of diffuse interfaces. Applications cover a wide range of fields, including oil and gas production, fluid transport in pipelines, environmental fluid dynamics, and materials science. The Navier-Stokes equations are often coupled with other models to simulate multiphysics phenomena, such as the Navier-Stokes-Allen-Cahn equations for fluid interfaces, and the Navier-Stokes-Magneto-hydrodynamic (MHD) equations for electromagnetic fluids.

Prior to the work of Navier and Stokes, the Euler equations, developed by Leonhard Euler in the 18th century, described ideal, i.e. non-viscous, fluid flows. These equations took no account of viscosity, a crucial factor in modeling real flows, and transformed fluid mechanics by providing a rigorous framework for modeling viscous flows. Their impact is immense and endures in many scientific and industrial fields, facilitating the understanding and prediction of complex fluid behavior. Navier, a French engineer and physicist, introduced a fluid flow model incorporating the effects of viscosity. He used solid mechanics concepts to include a viscosity term in Euler's equations, basing his approach on atomic theory. The Navier-Stokes equations, formulated by Claude-Louis Navier and George Gabriel Stokes in the 19th century, describe the motion of viscous fluids. They are widely used to model phenomena such as atmospheric currents, pipeline flow and industrial processes involving fluids. These equations are fundamental to fluid mechanics, governing the dynamic aspects of fluids under the influence of forces such as pressure, gravity and viscosity...

It's based on the free energy of the system and the competition between the different phases, leading to the formation of structures and interfaces within the material. In 1822, Navier published his work in "Mémoire sur les lois du mouvement des fluides", in which he proposed that viscous forces are proportional to rates of deformation in the fluid. Stokes, a British physicist and mathematician, independently developed similar equations. His work provided a more rigorous mathematical justification for Navier's equations and extended them to include more general boundary conditions. Stokes' contributions appear in "On the Theories of the Internal Friction of Fluids in Motion," where he systematically incorporates the effects of viscosity. The Navier-Stokes equations are gradually adopted by the scientific community to solve practical problems in hydrodynamics, such as flow in pipes and around submerged objects. They are applied to hydraulic engineering, canal and dam design, and other fields requiring an understanding of viscous flows. The advent of computers has made it possible to solve the Navier-Stokes equations numerically, paving the way for the simulation of complex fluid phenomena. Methods such as finite differences, finite elements and finite volumes are becoming standard tools in computational fluid dynamics (CFD). Modeling turbulence remains one of the major challenges in fluid mechanics. The Navier-Stokes equations provide the basis for turbulence models such as the Reynolds-averaged Navier-Stokes (RANS) equations and large-scale simulations (LES). As technology advances, the Navier-Stokes equations are being adapted to model flows at microscopic and nanoscopic scales, taking into account surface effects and complex boundary conditions. The Navier-Stokes equations are often coupled with other models to simulate multi-physical phenomena, such as the Navier-Stokes-Allen-Cahn equations for fluid interfaces, and the Navier-Stokes-Magnetohydrodynamic (MHD) equations for electromagnetic fluids.

The Allen-Cahn equations play an essential role in modeling phase transitions and interface dynamics in physical systems. Prior to the work of Allen and Cahn, phase transitions and interface dynamics were studied mainly through phenomenological models and experimental approaches. John W. Cahn, an American physicist and chemist, and Sam Allen, an American physicist, introduced their model in 1977 to describe interface dynamics in binary systems, where two different phases can coexist and evolve.

Their work was published in the article "A microscopic theory for antiphase boundary motion and its application to antiphase domain coarsening," in Acta Metallurgica. The article presented a mathematical model to describe how interfaces evolve over time. The Allen-Cahn equation models the competition between volume energy, which favors pure phases, and surface energy, which favors the creation of diffuse interfaces. The model includes terms representing surface tension and interface energy, thus capturing the dynamics of interface formation and movement. The Allen-Cahn model has been used to simulate coalescence and phase separation processes in materials. It can be used to predict the shape and movement of interfaces in multiphase systems. Research has demonstrated the model's effectiveness in studying domain coarsening, where small regions of phase coalesce to form larger regions over time. In parallel, Cahn also developed the Cahn-Hilliard model to describe phase separation and mass diffusion. Although similar to Allen-Cahn, the Cahn-Hilliard model includes a mass conservation term. Research continues to improve the theoretical understanding of the Allen-Cahn model, including stability analysis and solution dynamics. The model is used in various disciplines, such as materials science, chemical engineering, biophysics and environmental technologies, to study phenomena such as crystal growth, membrane dynamics, and chemical reactivity. In short, the Allen-Cahn equations, developed by John W. Cahn and Sam Allen, have revolutionized the way scientists model interface dynamics and phase transitions. Their elegant and efficient mathematical approach continues to serve as the basis for numerous research and industrial applications, offering profound insights into the complex dynamics of multiphase systems.

The coupling of the Navier-Stokes equations with the Allen-Cahn model represents a major advance in multi-phase fluid modeling, offering powerful tools for understanding and simulating complex interfacial dynamics. This integrated approach continues to play a key role in improving industrial processes and environmental systems. Introduced by John W. Cahn and Sam Allen in the 1970s, the Allen-Cahn model describes the evolution of a scalar order parameter representing a phase in a binary system. This model is mainly used to simulate phase transitions and interface dynamics. The Allen-Cahn equations play an essential role in modeling phase transitions and interface dynamics in physical systems. Here's a historical overview of their development and use. Prior to the work of

Allen and Cahn, phase transitions and interface dynamics were studied mainly through phenomenological models and experimental approaches. The study of binary mixtures of compressible, viscous and macroscopically immiscible fluids is of significant importance in various applications, such as oil and gas production, multiphase flow in pipelines, and environmental fluid dynamics. By gaining a better understanding of the behavior of these mixtures, it becomes possible to improve the design and performance of industrial processes and environmental systems. The Cahn-Hilliard-Navier-Stokes (CHNS) equations are also used to study the behavior of immiscible fluids in a variety of applications, including materials science, chemical engineering and phase separation processes .

Objective

The Allen-Cahn model is a mathematical approach used to model interface dynamics in multiphase systems, particularly in fluids where two distinct phases interact. The Allen-Cahn model defines the interface between two fluids as a smooth transition region, rather than a sharp, abrupt boundary. This region has extremely small but finite dimensions, where fluid properties gradually change from one phase to the other.

In a mixture of two immiscible fluids, such as oil and water, the interface according to the Allen-Cahn model is not a strict line of separation, but a zone where properties gradually change from those of oil to those of water. In this transition zone, the particles of the two fluids interact with each other. These interactions influence the fluid's local physical properties, such as density and viscosity. At the interface between oil and water, there is a mixing of particles from both fluids, creating a region where the characteristics are neither totally those of oil nor totally those of water. The physical properties of the fluid, such as density and viscosity, change continuously across this transition zone. This means that there is no abrupt jump in these properties; instead, they vary gradually with position in the domain. Although the physical properties change continuously, their exact values within the interface region are not precisely determined. This means that there is some ambiguity or imprecision as to their exact values at a given point in the interface. At the moment of the level-set formulation, which is another method for modeling interfaces, the Allen-Cahn model leaves certain key variables such as velocity, pressure and phase field undetermined. These variables must be solved simultaneously using the equations of fluid dynamics and appropriate boundary conditions. In fluid dynamics simulations, the Navier-Stokes equations could be coupled with the Allen-Cahn model to determine the velocity and pressure distribution as a function of the interface evolution described by the phase field. In summary, the Allen-Cahn model offers a sophisticated way of modeling interface dynamics in fluid systems by considering smooth, continuous transitions of physical properties influenced by the phase field, while accepting some indeterminacy of the exact values in the interface region.

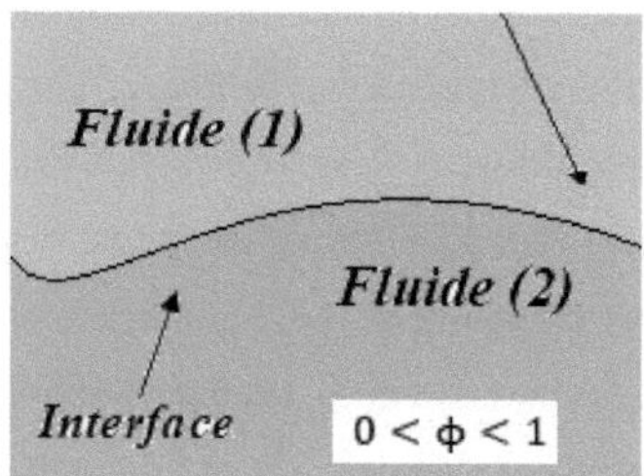

Fig. 1: Diagram of the interface between two incompressible fluids.

The Navier-Stokes equations describe fluid motion as a function of external forces such as pressure, gravity and viscosity. They are fundamental in fluid mechanics for understanding flow dynamics. The Navier-Stokes equations were formulated independently by Claude-Louis Navier and George Gabriel Stokes in the 19th century. They provide a detailed description of fluid dynamics in terms of the conservation of mass, momentum and energy. The Cahn-Hilliard equation was later developed to model phase separation processes in materials. It was introduced by John W. Cahn and John E. Hilliard in the 1950s and 1960s, in the context of metallurgy, to describe pattern formation during phase separation. The coupling of the Navier-Stokes equations with the Cahn-Hilliard equation makes it possible to model complex phenomena where fluid dynamics are influenced by phase separation. This is particularly useful for studying systems where two immiscible fluids interact, as in oil production processes, nuclear cooling systems, or environmental dynamics. CHNS are used to model multiphase flow in pipelines, helping to optimize hydrocarbon transport. They help to understand the interactions between different fluid layers in oceans and rivers, influencing the dispersion of pollutants. CHNS are used to study phase separation in alloys and polymers, helping to develop materials with specific properties. They are also used to model separation processes in chemical reactors and mixing systems. By combining the Navier-Stokes equations and the Cahn-Hilliard equation, CHNS offer a powerful framework for studying and simulating phenomena where fluid dynamics and phase separation play crucial roles.

I. Mathematical model

II.1 Mathematical Model of Two-Phase System Dynamics

Consider a system made up of two uniformly mixed fluids. Will the fluids remain mixed or separate, as in the case of oil and water? A simple experiment could provide the answer, but so can the thermodynamics of diffusion. The physics of diffusion is divided into positive and negative diffusion. Positive diffusion occurs in the opposite direction to the concentration gradient, thereby reducing the gradient, while negative diffusion occurs in the same direction as the gradient. The equilibrium state for positive diffusion is a uniform mixture, while for negative diffusion the equilibrium state is a two-phase system separated by a fluid interface. These two cases are considered separately, with positive diffusion as the dominant transport method in one case, and negative diffusion in the other. In the case of positive diffusion only, Fick's law of diffusion is taken into account, in combination with the continuity equation. Fick's first law of diffusion can be stated as follows:

$$\boldsymbol{J} = -\, D\nabla \mathrm{c} \qquad \text{(I.1)}$$

Where (c) is the concentration, D is the diffusivity and ***(J) is*** the concentration diffusion flux. The continuity equation is :

$$\frac{\partial c}{\partial t} + \nabla.\vec{J} = 0 \qquad \text{(I.2)}$$

The combination of Fick's first law of diffusion and the continuity equation for fluids leads to Fick's second law:

$$\frac{\partial c}{\partial t} = D\nabla^2 c \qquad \text{(I.3)}$$

According to equation (I.3), a uniform, homogeneous mixture is expected to be always in steady state: in the absence of a spatial concentration gradient, no concentration variation will occur over time. However, in the case of phase separation, diffusion takes place against the concentration gradient, which is not in accordance with equation (I.3). Thus, the concentration gradient is not the only driving force behind diffusion: another force must be at play. In the case of negative diffusion, the driving force is the chemical potential gradient, as described by Cahn and Hilliard in 1958.

A new expression for equation (I.1) can then be derived:

$$\boldsymbol{J} = -\,M\nabla\mu \qquad \text{(I.4)}$$

Where (M) is particle mobility (analogous to diffusivity (D) and (μ) is the chemical potential. Using this new driving force with the continuity equation, we obtain an alternative and more general form of Fick's second law. This equation is known as the Cahn-Hilliard equation, introduced by Cahn and Hilliard in 1958.

$$\frac{\partial c}{\partial t} = M\nabla^2\mu \qquad \text{(I.5)}$$

The concentration-dependent energy density of this two-phase system is also derived by Cahn and Hilliard, in (Cahn and Hilliard, 1958):

$$f(c) = \varepsilon\frac{c^2}{2}(1-c)^2 \qquad \text{(I.6)}$$

Where (c) is the local concentration and (ε) is a parameter representing the interfacial energy between phases.

On a macroscopic scale, a fluid can be seen as a continuous medium made up of a set of particles that can interact with each other. From a mathematical point of view, this translates into the continuity of the physical properties that characterize the fluid: density, velocity and pressure. At constant temperature, we present the mathematical model used to describe the dynamics of an incompressible Newtonian fluid. This model is based on the principles of mechanics: conservation of mass, conservation of momentum and conservation of energy.

- The principle of conservation of mass: The variation of fluid mass in a given volume over time is equivalent to the mass flow entering through the boundary. The local mathematical expression of mass conservation is :

$$\frac{\partial \rho}{\partial t} + \nabla.(\rho u) = 0 \quad (I.7)$$

Where (ρ) is the density of the fluid, When the fluid is incompressible, i.e. its density does not vary with time, the mass conservation equation (1.1) simplifies to :

$$\boldsymbol{\nabla}.u = 0 \quad (I.8)$$

Where ($\boldsymbol{u}$) is the velocity vector.

- Conservation of momentum: stipulates that the variation in momentum is equal to the sum of the external forces acting on the system, in accordance with Newton's second law. For a fluid, this principle is expressed by the following partial differential equation:

$$\rho\left(\frac{\partial \boldsymbol{u}}{\partial t} + +(\boldsymbol{u}.\boldsymbol{c})\boldsymbol{u}\right) = \boldsymbol{\nabla}.T(\boldsymbol{u},\boldsymbol{p}) + \rho\boldsymbol{f} \quad (I.9)$$

Where $\boldsymbol{T} = \boldsymbol{2\mu D(u) - pI}$ is the stress tensor and ($\boldsymbol{f}$) an external force applied to the system.

$$u.\boldsymbol{n} = 0 \quad (I.10)$$

In addition, ($\boldsymbol{n}$) is the unit normal vector pointing outwards (∂Ω).

- Conservation of fluid energy: equivalent to the first principle of thermodynamics. It can be expressed locally as follows:

$$\rho\left(\frac{\partial \boldsymbol{E}}{\partial t} + \boldsymbol{\nabla E}.\boldsymbol{u}\right) - \boldsymbol{\nabla}.(T.\boldsymbol{u} + \boldsymbol{q}) = \boldsymbol{f}.\boldsymbol{u} + \boldsymbol{r} \quad (I.11)$$

Where (***E***) is the energy density of the system and ($\boldsymbol{r}$) an external heat source.

II.2 Physical coupling: Navier-Stokes and Allen-Cahn equations

The Navier-Stokes equations are commonly used to describe the dynamics of interfaces between incompressible, macroscopically immiscible Newtonian fluids with compatible densities and viscosities. These equations express fluid motion and the conservation of mass, momentum and energy. By including additional terms to represent surface tension at fluid interfaces, the Navier-Stokes equations offer a precise description of fluid behavior, including interface evolution, flow patterns, and the transport of momentum and energy. Numerical resolution of these equations requires the use of methods such as finite differences or finite volumes, with appropriate boundary conditions to fully describe the system. Thus, the Navier-Stokes equations provide an essential mathematical framework for modeling the motion of fluids such as liquids and gases, incorporating the laws of conservation of mass and momentum as well as the effects of pressure, viscosity and other physical phenomena. Their resolution enables scientists and engineers to simulate a wide range of fluidic phenomena in fields such as aerodynamics, fluid mechanics and heat transfer. In fluid mechanics, the partial differential equations that describe the flow of incompressible fluids are the Navier-Stokes equations. In fluid mechanics, the partial differential equations describing the flow of incompressible fluids are the Navier-Stokes equations. In addition, the model is governed by the incompressible Navier-Stokes equations:

$$\partial_t \rho + \boldsymbol{div}(\rho \boldsymbol{u}) = 0 \tag{I.12}$$

$$u_t + (\boldsymbol{u}.\boldsymbol{\nabla})u + \boldsymbol{\nabla p} = v \Delta u + \zeta \kappa \boldsymbol{\nabla \varphi} \tag{I.13}$$

$$k(\boldsymbol{\varphi})\big(u_\sigma - u_\chi\big) + v \partial_n u_\sigma = \zeta k(\boldsymbol{\varphi}) \boldsymbol{\nabla}_\sigma \tag{I.14}$$

Where $(\boldsymbol{p})$ is the pressure, (v) is the coefficient of viscosity, defined as the ratio of shear stress to shear rate in a fluid. In other words, it describes the extent to which a fluid opposes motion when subjected to a force. The viscosity of a fluid can be influenced by temperature, pressure and chemical composition. In addition, the viscosity coefficient of two incompressible fluids can be compared by measuring the resistance to flow of each fluid. In general, a fluid with a higher viscosity coefficient will exhibit greater resistance to flow than a fluid with a lower viscosity coefficient. The viscosity of a fluid can also vary with temperature, so the viscosity coefficients of two fluids can change depending on the temperature at which they are measured. On the other hand, $(\boldsymbol{\varphi})$ represents the phase field variable and (ζ) is the capillary force in relation to the stress of a Newtonian fluid. In most practical cases, the capillary force is much weaker than the stress of the Newtonian fluid. Capillary force is proportional to fluid surface tension, while fluid stress is proportional to fluid pressure. For most liquids, surface tension is much lower than fluid pressure, resulting in a capillary force much lower than fluid stress.

In physical modeling, $k(\boldsymbol{\varphi})$ interfacial boundary energy plays an important role in many physical, chemical and biological processes, including liquid behavior, droplet and bubble formation, liquid propagation on solid surfaces, and interface stability in colloidal systems. Understanding and controlling interfacial boundary energy is crucial for applications such as materials synthesis, chemical separation and micro-fluidic device design. Boundary interfacial energy, also known as surface tension, is a measure of the energy required to create a new surface or interface between two phases. It represents the energy required to break the bonds between the molecules of one phase and create a new surface between the two phases. In a system containing several phases, the limiting interfacial energy is a measure of the energy required to increase the surface area between the phases. Finally, the (u_σ) is the limiting fluid velocity in the tangential direction, (u_χ) is the velocity of the boundary wall. We can define the vector operator $\boldsymbol{\nabla}_\sigma = (\boldsymbol{\nabla} - (n.\boldsymbol{\nabla})n)$ is the gradient in the tangential direction.

The interaction parameter at the interface between two fluids is often described by the surface tension coefficient, which represents the energy required to increase the surface area of the fluid-fluid interface. Surface tension results from the difference in intermolecular forces between the fluid molecules at the surface and those present in the fluid mass. The value of the surface tension coefficient depends on fluid properties such as density, viscosity and temperature. In a diffuse interface model, the surface tension coefficient is a key parameter affecting fluid behavior and interface dynamics. It determines the magnitude of the forces driving fluid movement and interface stability. Accurate determination of the surface tension coefficient is crucial for modeling the behavior of two-fluid systems subject to phase separation. Therefore, we consider the following system modeling the flow of a viscous, immiscible and incompressible two-phase fluid (1 and 2) with phase function :

$$\boldsymbol{\varphi}(x,t) = \begin{cases} 1 & \text{fluide 1} \\ -1 & \text{fluide 2} \end{cases} \tag{I.15}$$

In addition, (κ) is the chemical potential, or a measure of the energy per unit of substance required to add or remove a small amount of that substance to or from a system, while keeping its temperature and pressure constant. In this case, the chemical potential (κ) of each fluid is determined by its own temperature, pressure and chemical composition, and is independent of the presence of the other fluid. The two fluids will tend to mix until their chemical potentials are equal, at which point they will be in a state of chemical equilibrium. In this section, a mathematical model for such a system is given by the Cahn-Hilliard equations.

$$\kappa = -\xi \Delta\boldsymbol{\varphi} + f(\boldsymbol{\varphi}) \tag{I.16}$$

$$\boldsymbol{\varphi}_t + \boldsymbol{\nabla}.(\boldsymbol{u\varphi}) = \delta\Delta\kappa \tag{I.17}$$

$$\delta_t u - \Delta u + \Upsilon k(\boldsymbol{\varphi}) = 0; \quad \text{in } (0,t) \times \Omega \tag{I.18}$$

$$\partial_n u = 0; \text{ on } (0, t) \times (\partial\Omega) \text{ and } (x,t) \in \partial\Omega \times (0,+\infty) \tag{I.19}$$

Where (ξ) is the interface thickness and (Υ) is a boundary relaxation coefficient. In addition, we consider a diffuse interface model that describes the motion of an isothermal mixture of two immiscible and incompressible fluids undergoing phase separation, which can be described by the Navier-Stokes equations coupled to the Cahn-Hilliard equation.

$$\boldsymbol{\nabla}.u = 0 \tag{I.20}$$

$$\frac{\partial\varphi}{\partial t} + (\boldsymbol{u}.\boldsymbol{\nabla\varphi}) = \Upsilon\big(\alpha\Delta\boldsymbol{\varphi} - \boldsymbol{\beta}(\boldsymbol{\varphi}^2 - \boldsymbol{\varphi})\big) \tag{I.21}$$

$$\rho\frac{\partial(\rho u)}{\partial t} + \frac{1}{2}\boldsymbol{\nabla}.(\boldsymbol{\rho u})\boldsymbol{u} + (\boldsymbol{\rho u}.\boldsymbol{\nabla})\boldsymbol{u} - \boldsymbol{\nabla}.\big(\mu D(\boldsymbol{u})\big) + \boldsymbol{\nabla p} = \boldsymbol{f} - \frac{1}{\Upsilon}\left(\frac{\partial\varphi}{\partial t} + (\boldsymbol{u}.\boldsymbol{\nabla\varphi})\right)\boldsymbol{\nabla\varphi} \tag{I.22}$$

Where ($\boldsymbol{\varphi}$) is the phase field, ($\alpha = \beta/\varepsilon$^2) is a mixing energy density, (ε) is the transition zone thickness, (μ) is the viscosity, and D($\boldsymbol{u}$) is the shear rate:

$$D(\boldsymbol{u}) = \frac{1}{2}(\boldsymbol{\nabla u} + \boldsymbol{\nabla u}^T) \qquad \text{(I.23)}$$

And (Υ) acts as a relaxation of the time required to reach static equilibrium in the Allen-Cahn equation. On the other hand, the introduction of a surface tension term into the Navier-Stokes equations captures the effects of surface energy and the tendency of interfaces to minimize their surface area. The surface tension term is often proportional to the curvature of the interface, penalizing configurations with high curvature. This term encourages the system to reach a state with lower total surface energy. The coupling of these equations is common in studies related to fluid dynamics with phase transitions, such as in modeling the behavior of binary fluids undergoing phase separation. The additional elastic stress resulting from surface tension is defined as :

$$ST(\boldsymbol{\varphi}) = \mathcal{F}\boldsymbol{\nabla\varphi} \otimes \boldsymbol{\nabla\varphi} \qquad \text{(I.24)}$$

The constant $(\mathcal{F})$ represents the surface tension coefficient. The coupling between the Allen-Cahn model and the Navier-Stokes equations is often achieved through the stress tensor. The latter is the sum of the viscous stress tensor (σ^{Γ}) and the additional elastic stress term : $\sigma^{\Gamma} = \sigma_{\varepsilon}^{\Gamma} + ST(\boldsymbol{\varphi})$. This term captures the influence of surface tension on fluid flow, where variations in the order parameter φ are taken into account. Let's examine the Navier-Stokes equations applied to incompressible flow incorporating surface tension (measured in newtons per meter, $(\mathrm{N}m^{-1})$).

$$\partial_t \rho + \boldsymbol{u}.\boldsymbol{\nabla}\rho = 0 \qquad \text{(I.25)}$$

$$\boldsymbol{\rho(\partial_t u + u.\nabla u) = \nabla.[\,\mu(\nabla u + \nabla^T u)]\text{-}\,\nabla p + ST(\varphi)} \qquad \text{(I.26)}$$

An effective method for establishing an expression for $(\boldsymbol{ST}(\boldsymbol{\varphi})$involves examining the forces exerted on a two-dimensional curve under tension, a concept introduced by Young in 1805. Young's work in 1805 involved considering the forces acting on a two-dimensional curve under tension. In the context of fluid interfaces (excluding thin membranes), the application of two-dimensional tension is represented as a force per unit length tangential to the curve, denoted by (σt). Here, (t) represents the unit tangent vector and (σ) designates the surface tension coefficient. When considering an infinitesimal volume (Ω), intersected by the curve at points A and B, the total tension force acting on (Ω) can be expressed as follows:

$$\oint \boldsymbol{ST(\varphi)} = \int_A^B \boldsymbol{\sigma dt} = (\boldsymbol{\sigma_B t_B} - \boldsymbol{\sigma_A t_A}) \tag{I.27}$$

When a drop of liquid is deposited on a smooth solid surface, it forms a stable contact angle (θ) with the solid. This angle is closely related to the interfacial tensions between solid and liquid, solid and vapor, and liquid and vapor, as described by Young's equation :

$$\boldsymbol{ST(\varphi)_{SV}} = \boldsymbol{ST(\varphi)_{SL}} + \boldsymbol{ST(\varphi)_{LV}} \cos(\theta) \tag{I.28}$$

$\boldsymbol{ST(\varphi)_{SV}}$, $\boldsymbol{ST(\varphi)_{SL}}$ and $\boldsymbol{ST(\varphi)_{LV}}$ are respectively the interfacial forces per unit length between solid-vapor, solid-liquid and liquid-vapor at the contact line, i.e. the surface tension, and ($\theta = \widehat{\overrightarrow{\boldsymbol{ST(\varphi)_{SL}}}, \overrightarrow{\boldsymbol{ST(\varphi)_{SL}}}}$) is the contact angle (Fig. 2).

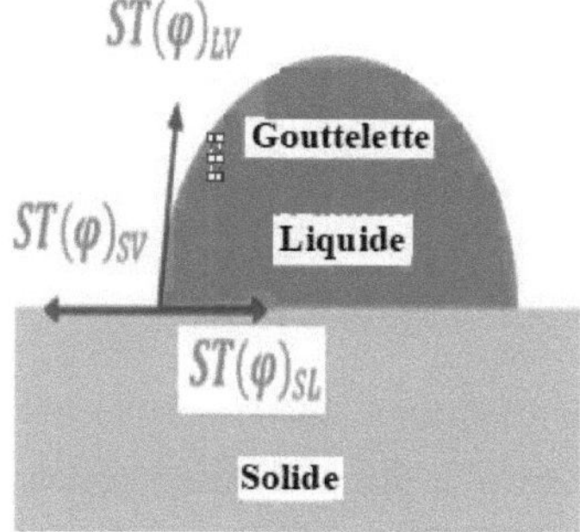

Fig. 2. Young's construction of the force equilibrium at a three-phase contact line.

The Navier-Stokes equation is used to model fluid behavior in a variety of applications, including aircraft design, fluid flow in pipelines, and the study of ocean currents, where :

$$\rho\left(\frac{\partial \boldsymbol{u}}{\partial t} + +(\boldsymbol{u}.\boldsymbol{c})\boldsymbol{u}\right) = \boldsymbol{\nabla}.T(\boldsymbol{u},\boldsymbol{p}) + \rho \boldsymbol{f} \tag{I.29}$$

As a result $\vec{T} = 2\mu D(\boldsymbol{u}) - pI$ is the stress tensor, (μ) is the dynamic viscosity and ($\boldsymbol{f}$) the external force. The static droplet is a fundamental two-phase problem, widely used to validate the numerical methods developed. This coupling leads to a set of equations describing the equilibrium state of a droplet at rest in a fluid. Here's the system of equations:

$$\frac{\partial u}{\partial t} - \boldsymbol{\nabla}.\left(\boldsymbol{M\nabla}\left(\frac{\partial F}{\partial u}\right)\right) = 0 \quad (I.30)$$

M is the mobility parameter, and F(u) is the double-well potential energy function. Furthermore, in Figure (2), we have quantitatively represented the density distributions along the horizontal centerline for different mobility values (M = 0.1; M = 2). The results show that the numerical predictions of the density field agree well with the analytical solution. At the interface, the concentration between two immiscible fluids depends on the properties of the two fluids and the forces acting on them. The interfacial energy between the two immiscible and incompressible fluids results from the differences in intermolecular forces between the two fluids. This represents the amount of energy available in a system to perform work at constant temperature and volume. In the case of two immiscible fluids, the Helmholtz free energy of the system is given by the sum of the Helmholtz free energies of the two individual fluids plus the interfacial energy between the two fluids. At the interface between the two fluids, the molecules are not in their preferred environment, resulting in an increase in the total free energy of the system. The Helmholtz free energy of the two immiscible, incompressible fluids is given by :

$$\frac{\partial \boldsymbol{\varphi}}{\partial \mathrm{t}} = \delta\left(\Delta\boldsymbol{\varphi} - \boldsymbol{f}(\boldsymbol{\varphi})\right) \quad (I.31)$$

With (δ) is the kinetic coefficient. The Helmholtz free energy of the system is minimized when the system is in thermodynamic equilibrium. In the case of two immiscible and incompressible fluids, this occurs when the interfacial energy is minimized and the interface adopts a shape that minimizes its surface area. On the other hand, the viscosity of a fluid mixture can be calculated using various methods, such as empirical models, mixing rules or molecular simulations. A commonly used model is the Krieger-Dougherty equation, which relates the viscosity of the mixture to the volume fraction and viscosity of each component, as well as to a parameter reflecting the degree of interaction between the molecules. In general, the viscosity of a mixture of two or more fluids is assumed to be higher than the viscosity of each of the individual components taken separately. This is because interactions between the molecules of different components can create additional resistance to flow, resulting in a higher overall viscosity. However, the exact behavior of the mixture depends on the specific properties of the fluids involved, and can vary according to factors such as temperature, pressure and shear rate. It depends on the viscosity of the individual components of the mixture, as well as their relative proportions and the interactions between the molecules of the different components. To describe the complete fluid dynamics, we define the viscosity of the fluid mixture as follows:

$$\mu(\boldsymbol{\varphi}) =: \mu^* + \zeta_\mu \boldsymbol{\varphi} \tag{I.32}$$

Or; $\mu^* = \frac{\mu_1+\mu_2}{2}$ and $\zeta_\mu = \frac{\mu_1-\mu_2}{2}$

Consequently, the density of a fluid mixture cannot always be accurately predicted using simple mixing rules, and may require more complex modeling or experimental measurements. The densities of individual components can be affected by temperature, pressure and other conditions, and may not be additive in all cases.

$$\rho(\boldsymbol{\varphi}) =: \rho^* + \zeta_\rho \boldsymbol{\varphi} \tag{I.33}$$

Or; $\rho^* = \frac{\rho_1+\rho_2}{2}$ and $\zeta_\rho = \frac{\rho_1-v}{2}$

In mathematics, in the field of differential equations, boundary conditions are practically essential for defining a problem, and are also of paramount importance in computational fluid dynamics.

$$u(0,x) = u_0 \tag{I.34}$$

$$\boldsymbol{\varphi}(0,x) = \boldsymbol{\varphi}_0 \tag{I.35}$$

With initial conditions :

$$u(x,0) = u_0(x) \tag{I.36}$$

$$\boldsymbol{\varphi}(x,0) = \boldsymbol{\varphi}_0(x) \tag{I.37}$$

Consequently, the total velocity at the interface (1 and 2) of two immiscible and incompressible fluids is :

$$\boldsymbol{u}^{Tot} = \boldsymbol{u}^{int(1)} + \boldsymbol{u}^{int(2)} \tag{I.38}$$

The Allen-Cahn equation can be derived by minimizing the energy functional under certain constraints, such as the conservation of mass or another physical quantity. The resulting equation describes the dynamics of the order parameter field (u) and is often used to model phase transitions and other phenomena in materials science, physics and various other fields. The energy functional is a measure of the total energy of the system and is frequently used to study the stability and dynamics of solutions of the Allen-Cahn equation. The Lyapunov energy functional associated with the Allen-Cahn equation is given by :

$$\oint \left(\frac{1}{2}|\boldsymbol{\nabla u}|^2 + \frac{1}{\varpi^2}\boldsymbol{H(u)}\right) \mathrm{dx} \tag{I.39}$$

The first term in the energy functional represents the kinetic energy of the system, while the second term represents the potential energy of the system. Where (ϖ) is a positive constant representing the interfacial energy between phases and $(\boldsymbol{\nabla u})$ is the gradient of (u) with respect to the spatial coordinates. Thus, $\boldsymbol{H(u)} = \frac{1}{4}(\boldsymbol{u}^2 - 1)^2$. The potential energy is minimized when the order parameter field takes the values (± 1), which correspond to the two stable phases of the system.

In general, there will be a difference in concentration between the two fluids due to the presence of surface tension at the interface. This difference can lead to the formation of a concentration gradient, with a higher concentration in the fluid with higher surface tension. The rate of mixing between the two fluids will also influence the concentration at the interface, as will any external forces acting on the fluids, such as convection, diffusion or agitation. As a result, we obtain the transport-diffusion equation describing the local change in concentration:

$$\partial_t c - D_0 \Delta c + \vec{u}.\boldsymbol{\nabla c} = 0 \tag{I.40}$$

Where $(D_0)$is the constant diffusion coefficient. At the fluid-fluid interface, we have :

$$\nabla c.n = 0 \quad (I.41)$$

II.3. Mathematical approach

The Allen-Cahn model, by coupling the phase-field evolution equation with the Navier-Stokes equations, enables the dynamics of two-phase systems to be modeled accurately and efficiently. It takes into account continuous variations in physical properties across the interface, and offers a robust mathematical approach to the study of phase transitions and similar phenomena. The Allen-Cahn model defines the interface as a smooth transition zone of extremely small dimensions, where distinct fluid particles interact. This approach implies only that the physical properties of the fluid, such as density and viscosity, change continuously across the entire domain. These changes are influenced by the phase field, a parameter that describes the state of the system at each point in space. In other words, in this transition zone, density and viscosity are not constant but vary progressively, reflecting the mixing of fluids. However, it is important to note that the exact values of these physical properties within the interface region cannot be precisely determined. This indeterminacy is due to the diffuse nature of the interface, where the characteristics of the fluids are superimposed. As in the level-set formulation, three key variables remain undetermined in this approach: fluid velocity, pressure and phase field. Velocity and pressure describe motion and forces in the fluid respectively, while the phase field captures the state of the system in terms of fluid phases. To accurately model mixing dynamics, it is necessary to define expressions for the density and viscosity of the fluid mixture. These definitions enable us to understand how the physical properties vary with position and time, and are essential for solving the governing equations, such as the Navier-Stokes equations coupled with the Allen-Cahn equations. In summary, the Allen-Cahn model offers a way of representing the interface between two fluids in a continuous and smooth manner, taking into account variations in physical properties and their influence by the phase field, while recognizing the precise indeterminacy of these properties in the region of the interface. Expressions for the density and viscosity of the mixture are given by :

$$\rho(\varphi) = \left(\frac{\rho_1-\rho_2}{2}\right)\varphi + \left(\frac{\rho_1+\rho_2}{2}\right) \quad \text{(I.42)}$$

$$\mu(\varphi) = \left(\frac{\mu_1-\mu}{2}\right)\varphi + \left(\frac{\mu+\mu_2}{2}\right) \quad \text{(I.43)}$$

$(\boldsymbol{\rho_1})$is the density of fluid 1 (chosen as a scaling factor for density) and $(\boldsymbol{\mu_1})$ is the viscosity of fluid 1 (chosen as a scaling factor for viscosity). The Cahn-Hilliard/Allen-Cahn model presented, introduced by J. Cahn and A. Novick-Cohen, is described by the following partial differential equations:

$$\frac{\partial \boldsymbol{m}}{\partial t} = \boldsymbol{h}^2\Delta(f(m+n) + f(m-n) - \boldsymbol{h}^2\Delta m) \quad \text{(I.44)}$$

$$\frac{\partial n}{\partial t} = -f(m+n) + f(m-n) - \alpha n + \boldsymbol{h}^2\Delta n \quad \text{(I.45)}$$

Where mmm is the concentration of one of the components and is a conserved quantity, $(\boldsymbol{n})$ is an order parameter, $(\boldsymbol{h})$ is a parameter representing the lattice spacing, $(\boldsymbol{\alpha})$ is a parameter that reflects the position of the system in the phase diagram (it can be either positive or negative), and $(\boldsymbol{f})$ is a function; the derivative of a double-well potential $(\boldsymbol{F})$is a function, and in this context is evaluated in $(\boldsymbol{m+n})$ and $(\boldsymbol{m-n})$. To demonstrate that the expression given for the free energy ***E(m, n)*** is indeed the integral of the specified terms, we'll expand the expressions and simplify where necessary. Starting with the expression given :

$$E(m,n) = \oint(F(m+n) + F(m-n) + \frac{\alpha}{2}n^2 + \frac{1}{2h^2}(|\boldsymbol{\nabla} m|^2 + |\boldsymbol{\nabla} n|^2))dx \quad \text{(I.46)}$$

Where (Ω) represents the integration domain and (f=F′) (f = F') (f=F′). The free energy associated with the Cahn-Hilliard /Allen-Cahn model given, we'll start with the expression of the free energy density.

$$F(m,n) = \int_0^{m,n} f(s)ds + \int_0^{m,n} f(s)ds \quad \text{(I.47)}$$

Now, let's substitute the expression of

$$E(m,n) = \oint [\int_0^{m,n} f(s)ds + \int_0^{m,n} f(s)ds + \frac{\alpha}{2}n^2 + \frac{1}{2h^2}(|\boldsymbol{\nabla} m|^2 + |\boldsymbol{\nabla} n|^2)]dx \quad (I.48)$$

Since $(f = F')$the integrals involving (f) can be expressed in terms of the potential (F) Let's analyze this step by step:

$$F(m+n) + F(m-n) = F(m) + F(n) + F(-m) + F(-n) \quad (I.49)$$

Combine similar terms:

$$F(m+n) + F(m-n) = F(m) + F(-m) + F(n) + F(-n) \quad (I.50)$$

Use the fact that: $(f = F')$the integrals involving (f) can be expressed as the potential(F):

$$E(m,n) = \oint (2F(m+n) + 2F(m-n) + \frac{\alpha}{2}n^2 + \frac{1}{2h^2}(|\boldsymbol{\nabla} m|^2 + |\boldsymbol{\nabla} n|^2))dx \quad (I.51)$$

$\boldsymbol{E(m,n)}$ Associated with the Cahn-Hilliard/Allen-Cahn model. In fluid mechanics and thermodynamics, the principle of energy conservation is a fundamental concept. It asserts that the total energy of a fluid system remains constant when subjected exclusively to conservative forces, such as gravity and pressure. This essential principle is rooted in the law of conservation of energy, a fundamental principle in physics.

$$E(\varphi,\mu) = \oint \left(\frac{\rho}{2}|u|^2 + \frac{\alpha}{2}|\nabla\varphi|^2 + \beta f(\varphi)\right) d\Omega \quad (I.52)$$

(φ) is a scalar field representing the concentration or phase of one of the fluids, (u) is the fluid velocity vector, (ρ) is the fluid density, (α) and (β) are constants governing the energy contributions of the(φ) and the deviation of(φ) from 1, respectively. Now, let's replace the expressions for $\rho(\varphi)$ and $\mu(\varphi)$ in the energy equation:

$$E(\varphi,\mu) = \frac{1}{2}\left(\frac{\rho_1-\rho_2}{2}\right)\varphi + \left(\frac{\rho_1+\rho_2}{2}\right)(u)^2 + \frac{\alpha}{2}|\nabla\varphi|^2 + \frac{\beta}{4}(\varphi-1)^2 \quad (I.53)$$

This expression quantifies the total energy of the system, taking into account both the kinetic energy of the fluid $(\frac{\rho}{2}|u|^2)$ and the energy associated with fluid mixing, which is influenced by the concentration gradient $\frac{\alpha}{2}|\nabla\varphi|^2$ and the deviation of (φ) from

$\left(\frac{\beta}{4}(\varphi-1)^2\right)$. Consequently, the total energy associated with the fluid mixture is given by the sum of these terms :

$$E(\varphi,\mu)=\frac{1}{4}(\rho_1-\rho_2)\varphi u^2+\frac{1}{4}(\rho_1+\rho_2)u^2+\frac{\alpha}{2}|\nabla\varphi|^2+\frac{\beta}{4}(\varphi-1)^2 \quad (I.54)$$

The mathematical expression used to represent conservation of mass is :

$$\frac{\partial(\rho V)}{\partial t}+\nabla.(\rho u)=0 \quad (I.55)$$

Where :$\left(\frac{\partial(\rho V)}{\partial t}\right)$: represents the rate of mass change within the system over time. $\nabla.(\rho u)$ Represents the divergence of mass flow, which is the rate at which mass enters or leaves a given region. In the case of a fluid, momentum equilibrium is described by the partial differential equation below:

$$\rho\left(\frac{\partial u}{\partial t}+(u.\nabla)u\right)=\nabla.T(u,p)+\rho f \quad (I.56)$$

Where $T(u,p)$is the stress tensor and (f) is an external force applied to the system.

The evolution of an isothermal mixture of two immiscible and incompressible fluids in a bounded domain can be accurately described by the Cahn-Hilliard-Navier-Stokes equations (CHNS system). These equations effectively model the dynamics of a two-phase fluid system by integrating both the fluid velocity and the concentration of one of the components. A common approach to modeling multi-fluid flows involving phase transitions is to couple the Allen-Cahn model with the Navier-Stokes equations. The Allen-Cahn equation captures the evolution of the phase field, while the Navier-Stokes equations describe the fluid flow. By coupling these two equations, we can simulate the evolution of fluid flow and the phase transition occurring at the interface between two phases. In addition, coupling can be achieved in different ways, depending on the specific problem and the desired level of accuracy. A common approach is to incorporate the phase field into the Navier-Stokes equation as a source term, taking into account density and viscosity changes at the interface. This allows complex phase interactions to be modeled in greater detail. Another approach is to use a level-set method, which offers a more accurate representation of the interface and fluid flow. The level-set method is particularly useful for tracking the position and shape of the interface with high accuracy, which is essential for simulations where interface details play a crucial role. In summary, modeling the evolution of an isothermal mixture of two immiscible and incompressible fluids can be efficiently achieved using the Cahn-Hilliard-Navier-Stokes equations, with approaches such as Allen-Cahn/Navier-Stokes coupling or the level-set method to capture the complex dynamics of two-phase fluids and phase transitions at the interface.

In addition, the coupling of the Allen-Cahn and Navier-Stokes equations leads to a strongly coupled, nonlinear system of partial differential equations, which presents significant challenges in terms of numerical solution. To solve this system, it is crucial to choose appropriate numerical methods. These methods must be capable of handling the complexity and non-linearity of the equations, while guaranteeing the stability and accuracy of the solutions. In addition, it is essential to correctly specify the boundary and initial conditions to obtain realistic and consistent simulations. The integrated model, which combines the Allen-Cahn model with the Navier-Stokes equations to simulate multi-fluid flows, does however have certain limitations. For example, it may not be suitable for accurately capturing very small-scale phenomena, such as micro-fluidic flows. In such cases, additional physical effects such as surface roughness or molecular interactions become significant and may require more complex modeling. In addition, this model often assumes isotropic surface tension, which may not accurately represent anisotropic surface tension effects in certain scenarios. This includes flows involving surfactants or complex fluid-fluid interfaces, where surface tension can vary depending on the direction and local conditions of the interface. As far as the Navier-Stokes equations are concerned, they are generally formulated for laminar flows. Consequently, they may not accurately depict turbulent flows, particularly at high Reynolds numbers. To model turbulent regimes correctly, it is often necessary to introduce turbulence models or make specific modifications to the equations to capture the effects of turbulence adequately. Finally, the model assumes that surface tension effects are adequately described by the Allen-Cahn equation. However, for highly dynamic or complex interfaces, this assumption may be insufficient. In such cases, it may be necessary to take into account additional surface tension effects to accurately represent interface behavior. Although the coupling of the Allen-Cahn and Navier-Stokes equations offers a powerful method for simulating multi-fluid flows, it is crucial to recognize and address its specific limitations. This includes adapting numerical methods, taking into account relevant scales, and incorporating additional models where necessary to ensure accurate and reliable predictions.

The Allen-Cahn model uses a single order parameter to describe phase separation. While this simplifies the modeling, it can also limit its ability to accurately represent complex multiphase systems with more varied phase behaviors. In other words, this model may not capture all the details of phase transitions in systems where phase interactions are more complicated than those described by a single parameter.

Furthermore, the Allen-Cahn model is based on the assumption that the system tends towards a uniform state over time, i.e. that it reaches equilibrium. This assumption may not be valid in all cases, especially for systems where non-equilibrium dynamics play an important role. Consequently, this simplification may lead to incorrect predictions in certain situations. Choosing appropriate boundary conditions, especially at fluid-fluid interfaces, can be complex and has a significant impact on model results. Boundary conditions must be carefully defined to correctly reflect interactions at the interface, as inappropriate choices can introduce significant errors into simulations. The model is also sensitive to the values of the various parameters used (parameter sensitivity). This sensitivity means that small variations in the parameters can lead to significant changes in the model results. Consequently, obtaining accurate experimental data to calibrate these parameters is essential, but often difficult. This calibration difficulty can limit the accuracy of model predictions. In practice, it is crucial to carefully consider the specific characteristics of the system to be modeled, and to assess the suitability of the Allen-Cahn model for the phenomena it intends to simulate. To ensure that the model is suitable for a particular application, it is important to validate its predictions against available experimental data. This validation verifies that the model faithfully reproduces the behavior observed in reality. Sensitivity analyses are also essential. These analyses study how parameter variations influence model results. They enable us to identify the most critical parameters and assess the robustness of the model's predictions in the face of uncertainties. In short, to use the Allen-Cahn model reliably, it is essential to carry out rigorous validation and in-depth sensitivity analyses, taking into account the specific characteristics of the system under study. The Allen-Cahn model uses a single order parameter to describe phase separation. While this simplification facilitates modeling, it can also limit the model's ability to accurately represent complex multiphase systems with more varied phase behaviors. In other words, the model may not capture all the details of phase transitions in systems where phase interactions are more complicated than those described by a single parameter.

Furthermore, the Allen-Cahn model is based on the assumption that the system tends towards a uniform state over time, i.e. that it reaches equilibrium. This assumption may not be valid in all cases, especially for systems where non-equilibrium dynamics play an important role. Consequently, this simplification may lead to incorrect predictions in certain situations.

Choosing appropriate boundary conditions, particularly at fluid-fluid interfaces, can be complex and have a significant impact on model results. Boundary conditions must be carefully defined to correctly reflect interactions at the interface. Inappropriate choices can introduce significant errors into simulations, affecting the accuracy of predictions. The model is also sensitive to the values of the various parameters used. This sensitivity means that small variations in the parameters can lead to significant changes in the model results. Obtaining accurate experimental data to calibrate these parameters is therefore essential, but can often be difficult. This difficulty in calibration can limit the accuracy of model predictions. In practice, it is crucial to carefully consider the specific characteristics of the system to be modeled, and to assess the suitability of the Allen-Cahn model for the phenomena it intends to simulate. To ensure that the model is suitable for a particular application, it is important to validate its predictions against available experimental data. This validation verifies that the model faithfully reproduces the behavior observed in reality. Sensitivity analyses are also essential. These analyses study how parameter variations influence model results. They enable us to identify the most critical parameters and assess the robustness of the model's predictions in the face of uncertainties. In short, to use the Allen-Cahn model reliably, it is essential to carry out rigorous validation and in-depth sensitivity analyses, taking into account the specific characteristics of the system under study.

In this context, our focus is on situations where one phase of a fluid mixture (known as immiscible) contains ions that may form solid precipitates at the interface between the fluid and a solid surface. Such precipitation can occur as a result of changes in temperature, pressure or other physical or chemical conditions that affect the solubility of ions in the fluid phase. The occurrence of ionic precipitation at the fluid-solid interface can have a variety of consequences. These include changes in the electrical conductivity of the fluid, changes in the adhesion between the fluid and the solid surface, and the formation of deposits on the solid surface.

To model the process of ionic precipitation at the pore scale, it is necessary to take into account several essential factors such as mass conservation, momentum and the presence of dissolved ions in each fluid phase. To this end, mathematical equations are used to describe the transport of fluids and ions through porous media. Free boundaries, which separate the different fluid phases, can be dynamic, evolving over time as ion precipitation and dissolution occur. To model these free boundaries effectively, numerical methods are commonly used. These include the level set method and the phase field method. These techniques make it possible to follow the evolution of the interface between the different phases of the fluid with precision. Using these numerical models, it is possible to make predictions about changes in the distribution of fluids and ions within porous media. These models also provide valuable information on the mechanism and kinetics of ionic precipitation at the pore scale. For example, they can help to understand how ions move and precipitate as a function of local temperature and pressure conditions, as well as the influence of the chemical properties of the fluid and solid surface. Modeling the process of ionic precipitation in porous media requires a thorough understanding of fluid and ion dynamics. Numerical methods such as level sets and phase fields play a crucial role in accurately representing dynamic interfaces and predicting the complex behavior of ion precipitation and dissolution. These tools make it possible to simulate and analyze the impact of various environmental factors on the distribution and behavior of ions in multiphase systems.

One of the advantages of the phase field method, or diffuse interface method, is its ability to simulate interface dynamics on an unadjusted mesh. This means that the mesh does not need to be adapted to conform to the interface, which greatly simplifies the simulation process. Indeed, it is not necessary to explicitly follow the interface, making the method more flexible and efficient compared to other numerical methods that rely on an adjusted mesh, such as the level set method. The use of an unadjusted mesh enables the diffuse interface method to handle complex and dynamic interfacial problems, including those with moving or deformable interfaces, without requiring time-consuming mesh adaptation. This greatly simplifies the modeling of systems where interfaces continually change shape and position. As a result, this method is particularly attractive for simulating a wide range of problems involving moving or deformable interfaces. This flexibility and efficiency have led to widespread use of the phase field method in a variety of fields, including materials simulations, multi-fluid flows and ionic precipitation phenomena. The phase-field model incorporates a small parameter associated with the thickness of the interface layer. This parameter, while essential for representing the gradual transition between phases, can introduce rigidity into the equations. This rigidity requires the use of specialized numerical techniques to solve the equations accurately. The use of efficient numerical methods that preserve energy stability at the discrete level is crucial to obtaining an accurate and efficient solution of the phase field model. These methods must be able to handle the rigidity introduced by the small interface thickness parameter, while maintaining the stability and accuracy of the simulations. For example, techniques such as the finite element method, implicit or semi-implicit schemes can be used to solve the phase field equations stably and accurately. As a result, the phase-field method offers a flexible and efficient approach to simulating complex, dynamic interfaces on unadjusted meshes. Its use simplifies simulations while maintaining high accuracy, thanks to the integration of specialized numerical techniques to manage the rigidity of the equations and preserve the energetic stability of the solutions.

The phase variable in the Cahn-Hilliard equation is a fictitious representation used to model phase transitions. Different variants such as Cahn-Hilliard, Allen-Cahn or other dynamics can be used to describe the behavior of this variable. The specific choice of dynamics used has a crucial impact on the accuracy and results of simulations. Thus, obtaining efficient and accurate solutions of the Cahn-Hilliard and Allen-Cahn equations is essential for understanding and predicting the behavior of materials during phase transitions. This understanding is crucial for the development of new materials with specific properties and for optimizing existing materials and manufacturing processes. Consequently, researchers are actively engaged in exploring various numerical methods to solve these equations accurately and efficiently. The aim is to realistically capture the complex phenomena associated with phase transitions and exploit this knowledge for practical applications. The Cahn-Hilliard-Navier-Stokes equations deal with the evolution of two immiscible, incompressible and isothermal fluids in a confined domain. This system of equations combines the Cahn-Hilliard equation, which describes the evolution of the order parameter distinguishing the two fluids, with the Navier-Stokes equation, which models fluid flow. By integrating these two aspects, this system of partial differential equations makes it possible to model and understand the behavior of complex fluid systems, particularly phase separation and multi-fluid flows. The diffuse interface model is often used to describe such systems, treating the interface between fluids as a diffuse zone rather than a sharp boundary (Fig. 1). One of the main advantages of the diffuse interface model is its ability to efficiently represent interface dynamics on an unadjusted mesh. Unlike other numerical methods, which require a matched mesh to accurately track the interface, the diffuse interface method simplifies this requirement. This makes it possible to model complex and dynamic interfacial problems, including those with moving or deformable interfaces, without the need for time-consuming mesh adaptation. In conclusion, models based on the Cahn-Hilliard and Allen-Cahn equations, as well as on the Cahn-Hilliard-Navier-Stokes equations with the use of the diffuse interface model, represent powerful tools for studying a variety of physical and chemical phenomena. These models are essential for understanding phase transitions, fluid separations and other complex phenomena on both microscopic and macroscopic scales. Their application, however, requires careful attention to the selection of appropriate dynamics and the use of suitable numerical methods to ensure accurate and meaningful results.

The phase variable in the Cahn-Hilliard equation is a fictitious representation used to model phase transitions. Different variants such as Cahn-Hilliard, Allen-Cahn or other dynamics can be used to describe the behavior of this variable. The specific choice of dynamics used has a crucial impact on the accuracy and results of simulations. Obtaining efficient and accurate solutions of the Cahn-Hilliard and Allen-Cahn equations is therefore essential for understanding and predicting the behavior of materials during phase transitions. This understanding is crucial for the development of new materials with specific properties, and for optimizing existing materials and manufacturing processes. Researchers are therefore actively engaged in exploring various numerical methods to solve these equations accurately and efficiently. The aim is to realistically capture the complex phenomena associated with phase transitions and exploit this knowledge for practical applications. The Cahn-Hilliard-Navier-Stokes equations deal with the evolution of two immiscible, incompressible and isothermal fluids in a confined domain. This system of equations combines the Cahn-Hilliard equation, which describes the evolution of the order parameter distinguishing the two fluids, with the Navier-Stokes equation, which models the fluid flow. This integrated approach makes it possible to model and understand the behavior of complex fluid systems, including phase separation phenomena and multi-fluid flows. The diffuse interface model is often used to describe such systems, treating the interface between fluids as a diffuse zone rather than a sharp boundary. This provides a more realistic and efficient representation of interface dynamics on an unadjusted mesh. Unlike other numerical methods, which require an adapted mesh to follow the interface precisely, the diffuse interface method simplifies this requirement. As a result, it can be used to model complex, dynamic interfacial problems, including those with moving or deformable interfaces, without the need for time-consuming mesh adaptation. While models based on the Cahn-Hilliard and Allen-Cahn equations, as well as on the Cahn-Hilliard/Navier-Stokes equations with the use of the diffuse interface model, represent powerful tools for studying a variety of physical and chemical phenomena. These models are essential for understanding phase transitions, fluid separations and other complex phenomena at microscopic and macroscopic scales. However, their effective application requires a rigorous selection of appropriate dynamics and the use of suitable numerical methods to guarantee accurate and meaningful results. This integrated approach not only enables these phenomena to be accurately modeled, but also provides important insights for the development of new materials and the understanding of industrial processes.

Conclusion

The model described in this work (a conservative Allen-Cahn equation with a Lagrange multiplier) is an extension of the standard Allen-Cahn equation and the Navier-Stokes equation, adapted to study the dynamics of phase transitions in fluids. This equation is a modified version of the standard Allen-Cahn equation, incorporating conservation of mass. Usually, the Allen-Cahn equation describes the evolution of an order parameter (such as a concentration) that distinguishes two phases. The conservative version adds a formulation that ensures that the sum of the phase volume fractions remains constant. The space-time-dependent Lagrange Multiplier is a scalar function used to impose a fixed volume fraction constraint on the two phases in the fluid. It plays a crucial role in the model formulation, guaranteeing the proper conservation of phase volume fractions during the phase transition. This model is widely used in Computational Fluid Dynamics to study phase separation in multiphase flows. It enables analysis of how phase transitions affect overall fluid behavior, particularly with regard to phase distribution and transport properties. In this work, the focus is on a specific variant of this model, known as the Cahn-Hilliard/Navier-Stokes (CHNS) system. This system combines aspects of the phase dynamics described by the Cahn-Hilliard equation with the flow dynamics described by the Navier-Stokes equations. This approach enables integrated and accurate modeling of complex fluid systems undergoing phase transitions. The book aims to study this system theoretically through an innovative mathematical model based on the NS-AC (Navier-Stokes and Allen-Cahn) model as a coupled nonlinear system. This provides a better understanding of phase separation phenomena and valuable information on fluid behavior during phase transitions. In short, this model represents a significant advance in the modeling of multiphase flows and in the understanding of phase transition phenomena, by integrating key aspects of fluid dynamics and mixture thermodynamics.

Chapter II: Numerical approach to phase separation

Before Proposing

The study and numerical analysis of fluid interfaces represents a crucial discipline in computational fluid dynamics, aimed at understanding the complex behaviors of interfaces between different fluid media. This approach combines the principles of the Navier-Stokes equations to describe fluid motions with models such as the Allen-Cahn model to characterize phase transitions and interfaces. By integrating these models, numerical simulation can explore a variety of phenomena, such as phase separation, coalescence and droplet dynamics, at different spatial and temporal scales. This introduction focuses on the advanced methods and practical applications of numerical analysis of fluid interfaces, highlighting its essential role in scientific research, engineering and industrial process modeling.

The numerical approach to phase separation is an essential area of computational fluid dynamics, aimed at modeling and understanding the complex phenomena of component separation in multiphase systems. This approach relies on the use of advanced mathematical models and numerical methods to simulate the interactions between fluid phases and predict their dynamic behavior. Phase separation occurs when fluid mixtures, such as chemical solutions or molten metal alloys, spontaneously separate into two distinct phases due to physico-chemical differences such as temperature, composition or pressure. Understanding this process is crucial not only for industry, where it is used in the design of new materials and in manufacturing processes, but also for diverse fields such as biology, the environment and geology. Numerical studies of phase separation involve several key steps, including the formulation of mathematical models describing interfacial forces and fluid motions, the discretization of these models to suit numerical resolution on computer grids, and finally the rigorous validation of the results obtained by comparison with experimental data or analytical solutions.

This section presents a detailed exploration of the methods and models used to simulate and analyze the dynamics of fluid interfaces. It discusses the key stages of numerical modeling, from the mathematical formulation of phase separation equations to the validation of the results obtained by numerical methods. The main aim of this study is to provide an in-depth understanding of phase separation processes and demonstrate how numerical simulations can be used to predict the behavior of complex fluid systems. By presenting concrete examples and discussing the challenges and opportunities associated with this approach, we hope to offer researchers and engineers a useful guide to the application of numerical techniques in their own work on multiphase fluids.

I. Study and Numerical Analysis of Fluid Interfaces

In general, a diffuse interface system is a model that considers the interface between two phases of a material as a region where material properties change gradually over a certain thickness, rather than abruptly. This approach makes it possible to model phase transitions more realistically, by capturing gradual variations in material properties within the interface. The Allen-Cahn equation is a partial differential equation often used to describe the dynamics of phase separation in materials. It models the evolution of an order parameter, representing the concentration or density of one phase relative to another, taking into account the forces that tend to homogenize the system and the local interactions that favor the formation of distinct domains. The Navier-Stokes equations describe fluid motion, taking into account the conservation of mass, momentum and energy. They are fundamental to modeling fluid flows, whether incompressible or compressible, and incorporate effects such as viscosity and applied external forces. By coupling the Allen-Cahn equation with the Navier-Stokes equations in a system with a diffuse interface, the behavior of the material or fluid at the interface can be analyzed more realistically. This approach makes it possible to track the evolution of moving interfaces, model phase interactions and study the effects of fluid dynamics on phase separation. However, it is also important to study the net interface limit, where the thickness of the diffuse interface tends towards zero. In this case, the system becomes equivalent to a net interface model, where the interface is treated as a discontinuity between phases. This limit is useful for simplifying certain calculations and for comparing the results of diffuse interface models with analytical or numerical solutions obtained from net interface models. In summary, the diffuse interface approach enables more detailed and realistic modeling of phase transitions, while the study of the net interface boundary offers complementary perspectives and useful comparisons for validating and refining numerical models.

I.1 Net Interface Limit study

Studying the net interface boundary can be complex, as it requires careful analysis of the system's behavior as the interface thickness tends towards zero. This analysis can involve studying the behavior of the system near the interface using techniques such as asymptotic analysis. Alternatively, numerical simulations can be used to explore the behavior of the system for small interface thicknesses. By examining the net interface limit of a diffuse interface system that couples the Allen-Cahn equation with the Navier-Stokes equations, we can gain valuable insights into the behavior of complex materials and fluids at interfaces, enriching our understanding of various physical phenomena. This enables us to model more accurately the dynamics of phase transitions and multi-fluid flows, which are essential in many scientific and industrial fields. To solve the Navier-Stokes /Allen-Cahn system of equations, a combined approach using dimensional analysis and finite difference methods is adopted. The use of dimensional analysis enables the governing equations to be transformed into an adimensional form. This facilitates a more compact representation and reveals the non-dimensional parameters that govern the system's behavior. The original variables are divided by appropriate reference quantities to obtain dimensionless variables, thus simplifying the equations and making them more universal.

After dimensional analysis, finite-difference methods are applied to discretize the transformed equations in the spatial and temporal domains. This discretization makes it possible to approximate the continuous spatial and temporal variations of the system on a grid, decomposing the problem into a set of algebraic equations that can be solved numerically. Finite difference methods provide a practical numerical framework for solving the system and obtaining solutions that capture the dynamics of fluid flow and phase transition phenomena. By implementing this dual methodology, the system of Navier-Stokes/Allen-Cahn equations is efficiently transformed, simplified and made accessible to computational solution. Adimensional analysis helps to understand the underlying physics, while finite-difference methods enable the system to be solved numerically and complex dynamics to be captured. To simplify the Navier-Stokes/Allen-Cahn system for a mixture of two fluids, the introduction of adimensional variables and parameters is necessary. This process involves selecting appropriate reference quantities for the system of interest and dividing the original variables by these references. Here are the scales we will use to obtain adimensional numbers:

$$u^* = \left(\frac{u}{U_0}\right), \mu^* = \left(\frac{\mu}{\mu_0}\right), \rho^* = \left(\frac{\rho}{\rho_0}\right), p^* = \left(\frac{p}{P_0}\right), \quad x^* = \left(\frac{x}{L}\right), \quad t^* = \left(\frac{U_0 t}{L}\right), \tag{II.1}$$

Where (U_0) is a characteristic velocity of the problem, *(L)* is a characteristic geometric dimension. By adopting this approach, we can not only simplify the equations, but also reveal the key parameters that influence the system's behavior, facilitating further analysis and understanding of phase separation phenomena in fluids. A g(t,x) function is a relationship that assigns a single value (g_0):

$$g_0 g^*(t^*, x^*) = \left(\mathbf{g}\left(\frac{U_0}{L}\right)t^*, (Lx^*)\right), \nabla^* g^*(t^*, x^*) = \left(\frac{U_0}{L}\right)\nabla \mathrm{g}(\mathrm{t}, \mathrm{x}) \tag{II.2}$$

We can have:

$$\nabla^*. u^* = 0 \tag{II.3}$$

$$\left(\rho_0 {U_0}^2 \rho^*(\boldsymbol{\varphi})\right)\left(\frac{\partial u^*}{\partial t^*} + (u^*.\nabla^*)u^*\right) - \frac{\mu_0 U_0}{L}\nabla^*.\left(2\mu^*(\boldsymbol{\varphi})\right)\boldsymbol{D}^*(u^*) + \frac{\mu_0 U_0}{L}\nabla^* p^* = \rho_0 \chi L \rho^*(\boldsymbol{\varphi}) - \frac{U_0}{\tau L}\left(\frac{\partial \varphi}{\partial t^*} + (u^*.\nabla^* \boldsymbol{\varphi})\right)\nabla^* \boldsymbol{\varphi} \tag{II.4}$$

$$U_0\left(\frac{\partial \varphi}{\partial t^*} + (u^*.\nabla^* \boldsymbol{\varphi})\right) = \tau L\left(\frac{\eta}{L^2}\Delta^* \boldsymbol{\varphi} - \boldsymbol{\beta}(\boldsymbol{\varphi}^2 - \boldsymbol{\varphi})\right) \tag{II.5}$$

With N_F is the Froude number, N_C is the Cahn number and R_e is the Reynolds number. Once these quantities have been identified, the dimensional analysis equations can be obtained:

$$N_F = \left(\frac{{U_0}^2}{\chi L}\right), N_C = \left(\left(\frac{\eta}{\boldsymbol{\beta}}\right)^{\left(\frac{1}{2}\right)}/L\right), R_e = \left(\frac{\rho_0 U_0 L}{\mu_0}\right) \tag{II.6}$$

(β, χ) Are positive constants. In this section, we continue our discussion of dimensionality by looking at non-dimensional numbers. We'll briefly introduce dimensionless numbers at this point:

$$\nabla^*.u^* = 0 \tag{II.7}$$

$$(\rho^*(\boldsymbol{\varphi}))\left(\frac{\partial u^*}{\partial t^*} + (u^*.\nabla^*)u^*\right) - \frac{1}{R_e}\nabla^*.\left(2\mu^*(\boldsymbol{\varphi})\right)\boldsymbol{D}^*(u^*) + \frac{1}{R_e}\nabla^* p^* = \tag{II.8}$$

$$\frac{1}{N_F}\rho^*(\boldsymbol{\varphi}) - \frac{1}{(\tau\mu_0)R_e}\left(\frac{\partial\varphi}{\partial t^*} + (u^*.\nabla^*\boldsymbol{\varphi})\right)\nabla^*\boldsymbol{\varphi}$$

$$\frac{\partial\varphi}{\partial t^*} + (u^*.\nabla^*\boldsymbol{\varphi}) = -\frac{2\eta\tau}{U_0}\frac{1}{N_C}\left((N_C)^2\Delta^*\boldsymbol{\varphi} + \boldsymbol{\varphi}(\boldsymbol{\varphi}^2 - \mathbf{1})\right) \tag{II.9}$$

I.2 Numerical results

In order to obtain results, we present a numerical iterative method for approximating the solutions of the nonlinear Cahn-Hilliard-Navier-Stokes equations. This section includes numerical tests that demonstrate the accuracy and efficiency of this approach for approximating solutions of the Cahn-Hilliard-Navier-Stokes equations, by comparing it with other numerical experimental approaches. It is important to note that the results obtained from the Crank-Nicolson scheme for approximating the Allen-Cahn equation will depend on the specific parameters of the problem and the initial conditions chosen (Figs. 3, 4 and 5). Generally speaking, however, the Crank-Nicolson method is a stable and accurate scheme for solving diffusion equations such as the Allen-Cahn equation. An advantage of the Crank-Nicolson method is that it offers second-order accuracy in both time and space.

This means that numerical errors in the solution decrease as the time step size and grid spacing decrease. What's more, the method is unconditionally stable, allowing a time step size to be chosen that is independent of the diffusion coefficient and regularization parameter. In general, the results obtained with the Crank-Nicolson scheme for approximating the Allen-Cahn equation should be accurate and stable, provided that the problem parameters and initial conditions are chosen appropriately, and the time step size and grid spacing are small enough to meet the method's stability and accuracy requirements. We now present the numerical solution of the concentration evolution of a binary mixture using the Allen-Cahn equation (Fig. 6).

The Allen-Cahn equation is a mathematical model used to describe the evolution of a binary system composed of two distinct phases or states. The concentration field is a parameter that characterizes the distribution of these phases in the spatial domain of the system. In the context of the Allen-Cahn equation, the concentration field is a function that quantifies the difference in concentration between the two phases. Over time, the interface between the two phases becomes progressively smoother due to the diffusion term in the equation. This term acts to attenuate and reduce existing concentration gradients in the system. As a result, the concentration field becomes more uniform and homogeneous over time.

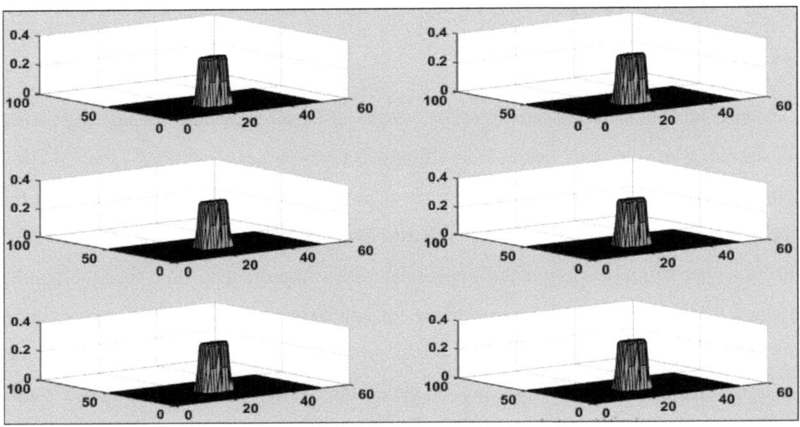

Fig. 3: Numerical solution of the Allen-Cahn equation.

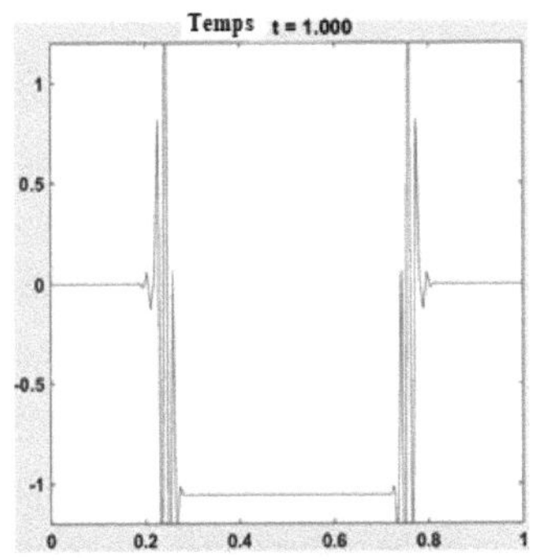

Fig. 4: Time evolution of the Allen-Cahn equation.

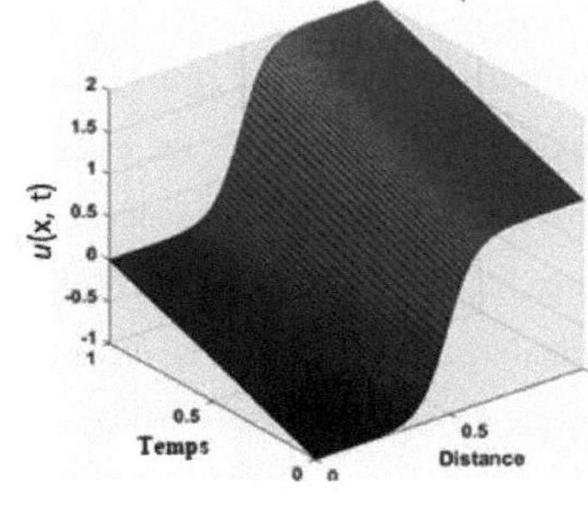

Fig. 5: Solution of the Allen-Cahn equation

$u(x, t)$

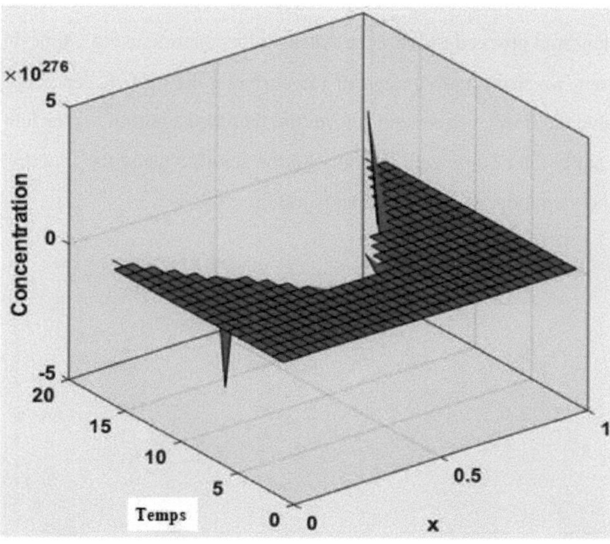

Fig. 6: Evolution of the concentration of a binary mixture using the Allen-Cahn equation.

In simple terms, the diffusion term in the Allen-Cahn equation has the effect of smoothing the interface between the two phases. This means that the boundary between the phases becomes less sharp and more gradual. To analyze the evolution of the concentration field in this equation, we use a partial differential equation, which can be solved numerically using a variety of methods. By solving this equation, we can make predictions about how the concentration field will evolve over time and how the interface between the two phases will transform. The concentration field plays a crucial role in understanding the dynamics of phase transitions and enables us to study a variety of physical phenomena, including crystal growth, patterning in fluids and the behavior of magnetic materials.

The finite difference method is used to obtain numerical results, which are then visualized using the surf function in (Fig. 7). In this implementation, we specify several parameters for the problem, including the domain length in the x and y directions, the final time, the number of grid points in each direction, the grid spacing, the time step size, the diffusivity parameter alpha, and the double-well potential parameter epsilon. Subsequently, we define the discrete Laplacian operator using finite-difference approximations and proceed to solve the Allen-Cahn equation using a time-step loop. At each time step, we first create a copy of the current solution (*u*), denoted by (*u_old*). Next, we solve the linear system and reform the flattened solution vector into a 2D grid. Finally, we apply the double-well potential to the solution using the function (*np.tanh*). After the time-step loop, we plot the final solution.

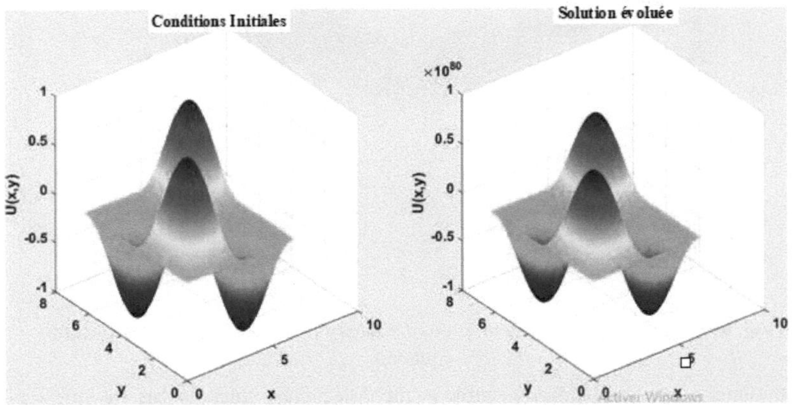

Fig. 7. 2D evolution solution of the Allen-Cahn equation (x, y).

The singular value decomposition (SVD) of the snapshot matrix, obtained from a collection of solution snapshots taken over time, allows us to identify the dominant modes of variability present in the solution. The singular values in the (SVD) indicate the energy content associated with each mode. A rapid decay of the singular values suggests that the majority of the solution's energy is concentrated in a few dominant modes. This implies that the solution can be well approximated by a lower-dimensional subspace formed by these dominant modes, rather than by the full-dimensional space encompassing all discretized variables. By projecting the solution onto this subspace, we can obtain an accurate approximation of the solution using a reduced set of variables. This forms the basis of various model reduction techniques aimed at identifying and projecting the solution onto the dominant modes. These techniques are used to reduce the computational complexity of simulations while maintaining accuracy. In the case of phase separation systems, the Allen-Cahn equation is a partial differential equation used to describe the evolution of a scalar order parameter. This order parameter can be interpreted as representing the concentration of a substance (Figs. 8 and 9).

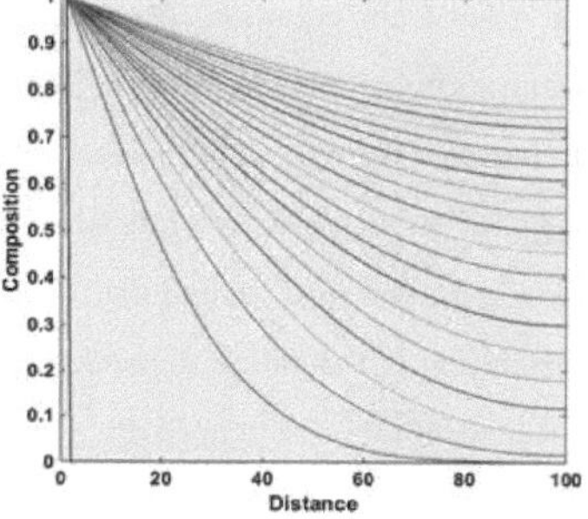

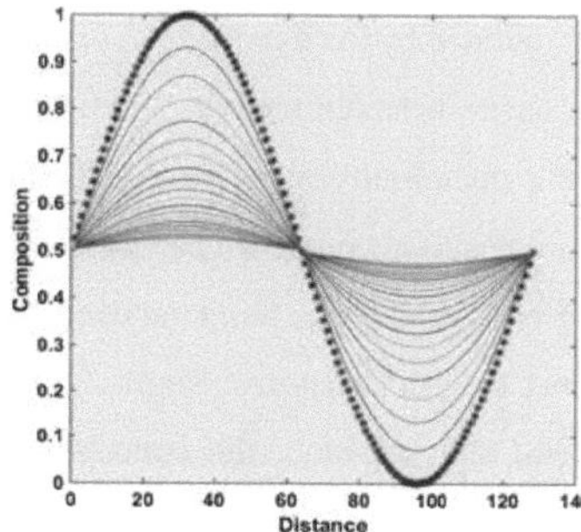

Fig. 8. surface concentration. **Fig. 9.** 1D concentration profile.

In a system undergoing phase separation, we can extend the Allen-Cahn equation to incorporate the effects of distance and composition. By modifying the Allen-Cahn equation, we can take into account the influence of distance on system behavior. The Allen-Cahn equation is a partial differential equation used to describe the evolution of a scalar parameter, representing the concentration of a substance. Similarly, in two-component systems, we can introduce a composition field to model the interactions between the different components of the system. This extension enables us to better understand and predict the complex dynamics of phase separation by considering both spatial variations and compositional differences. We can introduce a composition field c(x) which characterizes the local solute concentration. In both cases, an additional term is included in the equation to account for the impact of distance or composition on the evolution of the order parameter. This improved formulation provides a more realistic representation of phase separation in complex systems, where these factors play a significant role.

Incorporating distance and composition effects into the Allen-Cahn equation can lead to significant changes in the concentration profile of a system undergoing phase separation. When distance dependence is considered, the system tends to develop distinct regions characterized by different concentrations, separated by sharp interfaces. The distance function acts as a driving force for the formation of these interfaces. Regions closer to the boundary have higher solute or component concentrations, creating sharp concentration gradients. Over time, these interfaces will undergo movements and changes in shape dictated by the dynamics governed by the Allen-Cahn equation. This means that the boundaries between regions of different concentration can move, merge or separate, depending on initial conditions and system parameters. By taking both distance and composition into account, we can model phase separation phenomena more accurately, enabling us to predict the temporal evolution of concentration profiles and the formation of internal system structures. Similarly, by incorporating composition dependence into the Allen-Cahn equation, the concentration profile becomes influenced by the local concentration of the solute or component, as represented by the composition field c(x). This leads to the formation of regions of high and low concentration, separated by diffuse interfaces where the concentration gradually shifts from one value to another. Once again, the dynamics of the Allen-Cahn equation drive the evolution of the concentration profile over time. By including distance and composition effects, the Allen-Cahn equation offers a more realistic modeling approach for phase separation in complex systems, where these factors play a crucial role in determining the concentration profile.

In the case of two immiscible, incompressible fluids, (Fig. 10) illustrates how a star-shaped interface progresses in a flow influenced by curvature.

Fig. 10. 3D interface between two immiscible, incompressible fluids.

This interface develops as the two fluids interact, and is influenced by the curvature gradients present in the system. The temporal evolution of this interface between the two fluids is mathematically described by the Cahn-Hilliard equation. This partial differential equation captures how the scalar-order parameter field, which represents the local concentration or composition of the fluids, varies in space and time. The interface itself is visualized as an iso-surface of this scalar field, meaning that this surface represents the points where the value of the field is constant. In addition, the Cahn-Hilliard equation is formulated to describe the dynamics of phase separation in systems where two immiscible phases coexist. It takes into account the curvature forces that lead to the formation and evolution of complex interface structures, such as those observed in (Fig. 10).

This equation is crucial to understanding how the patterns and shapes of fluid interfaces evolve over time under the influence of surface forces and concentration gradients.

$$\frac{\partial \phi}{\partial t} = \mathrm{M}\,\nabla^2 \left(\frac{\Delta \mathrm{G}}{\Delta \phi}\right) - \varsigma\,(\nabla^4 \phi) \qquad \text{(II.10)}$$

Where (φ) is the order parameter, (M) is the mobility coefficient, ($\Delta G/\Delta\varphi$) is the free energy density, ς is the gradient energy coefficient. The solution to the equation provides the evolution of the interface between the two fluids over time. The resulting iso-surface can be visualized using a variety of techniques, including running cubes, ray-tracing and volume rendering. These visualization methods provide a detailed representation of the interface and its evolution throughout the simulation.

Spinodal decomposition can be modeled using the Cahn-Hilliard equation, which captures the dynamics of the order parameter in a system undergoing phase separation. The order parameter is a variable that indicates the extent of mixing between the two distinct phases of a material. It can take on different values to represent different phases, enabling the evolution of the material's structure to be followed over time. In spinodal decomposition, an initially homogeneous phase splits into two distinct phases due to thermodynamic instability. This phenomenon occurs when the system reaches a certain temperature and concentration condition, rendering the homogeneous phase unstable and favoring the formation of domains rich and poor in one of the components of the mixture. The Cahn-Hilliard equation is a partial differential equation that describes the dynamics of this spinodal decomposition. It takes into account both diffusion and near-field interactions between the components of the mixture. This equation models how the order parameter evolves in time and space, in response to concentration gradients and diffusion forces that tend to minimize the free energy of the system. The Cahn-Hilliard equation can be expressed as follows:

$$\frac{\partial \phi}{\partial t} = M\, \nabla^2(\phi^3 - \phi - \varsigma^2\, \nabla^2\phi) \tag{II.11}$$

The term ∇^2 (represents the Laplacian operator), which describes the spatial variation of the order parameter. To visualize the evolution of the iso-surface, we can use techniques such as contour plotting or surface plotting to represent the interface between the two phases (Fig. 11). The iso-surface represents the boundary between regions where the order parameter has positive and negative values, corresponding to the two distinct phases. As the system evolves, this boundary will become increasingly complex and fragmented, forming a pattern of interconnected domains characteristic of spinodal decomposition.

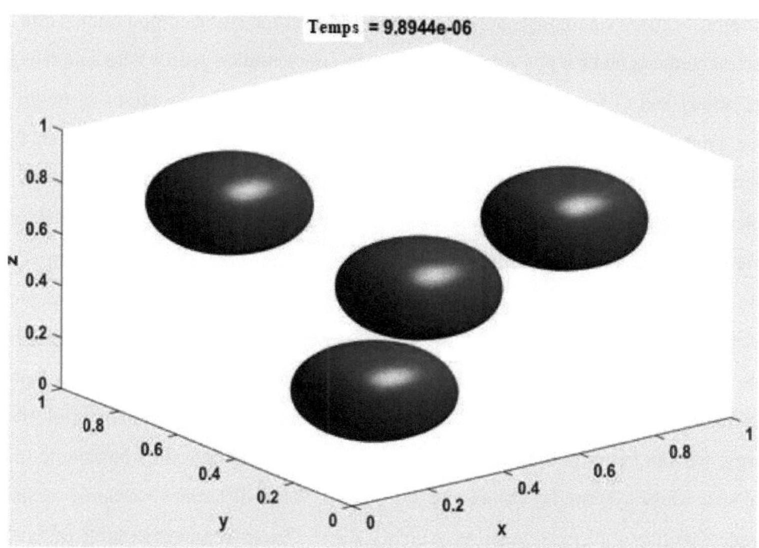

Fig. 11. 3D iso-surface between two immiscible and incompressible fluids.

The Allen-Cahn equation is a reaction-diffusion equation used to describe the dynamics of phase separation in binary alloys, where the concentration of a component varies in different regions of the material. This equation models the evolution of a component's concentration over time and space, incorporating both diffusion and reaction terms to capture the mechanisms underlying phase separation. The diffusion term in the Allen-Cahn equation describes the movement of atoms or molecules in response to concentration gradients, tending to homogenize the concentration distribution. The reaction term, on the other hand, represents local interactions between components, favoring the formation of distinct domains and phase separation. Together, these terms lead the system to a state of lower energy, where the components are separated into distinct phases with clear interfaces between them. To analyze this complex system, we employ two main numerical approaches: spectral and finite-difference methods. These methods allow us to solve the Allen-Cahn equation accurately and efficiently, offering complementary insights into the dynamics of phase separation. Spectral methods use series of orthogonal functions, such as Fourier series or Chebyshev polynomials, to represent the solution of the equation. This approach is particularly useful for periodic problems, or those where high accuracy is required over the entire computational domain. Spectral methods make it possible to calculate the concentration profile with a high degree of accuracy and to follow its temporal evolution by solving the equations in frequency space. Finite-difference methods, on the other hand, approximate the derivatives by finite differences on a discrete grid. This approach is more intuitive and often easier to implement for complex geometries or varied boundary conditions. Finite-difference methods allow us to follow the evolution of the concentration profile by solving the Allen-Cahn equation at each grid point over time. Interfacial energy, a measure of the energy associated with the presence of interfaces between separate phases, can also be calculated using these two techniques. This energy plays a crucial role in the dynamics of phase separation, influencing the shape and stability of the interfaces. By calculating interfacial energy, we can better understand how interfaces evolve and how they contribute to the final state of the system. By combining spectral and finite-difference methods, we gain a complete picture of phase separation in binary alloys. These techniques enable us to solve the Allen-Cahn equation with high accuracy, follow the evolution of the concentration profile over time and determine the interfacial energy, offering valuable insights into the mechanisms of phase separation and the properties of the resulting materials.

The finite-difference method (FDM) for calculating concentration is a widely-used numerical approach to solving the Allen-Cahn equation. It involves discretizing the equation on a regular grid, replacing partial derivatives with finite-difference approximations. This method transforms the continuous differential equation into a set of discrete algebraic equations that can be solved by numerical algorithms. The finite difference method has been successfully applied to a variety of physical and engineering problems involving phase separation and patterning. For example, in binary alloys, the Allen-Cahn equation can model the dynamics of phase separation where two initially mixed components separate into distinct domains over time. The patterns formed can represent micro- or nanometric structures, such as precipitates in metallic materials or phase patterns in polymers. To apply the finite-difference method, the Allen-Cahn equation is discretized in both time and space. The stability and accuracy of this method depend crucially on the selection of time (Δt) and space (Δx) steps. An appropriate choice of (Δt) and (Δx) is necessary to ensure that the numerical solution converges to the exact solution of the equation. The spatial region is divided into a regular grid, where each grid point represents a discrete position in space. The spatial derivatives in the Allen-Cahn equation are approximated by centered or off-center finite differences. Time is also discretized into regular intervals. Time derivatives are approximated using schemes such as forward, backward or centered differences. A common scheme is the explicit Euler scheme. The stability and accuracy of the finite-difference method for calculating the concentration in the Allen-Cahn equation are strongly influenced by the choice of time and space steps. A common stability criterion is the Courant-Friedrichs-Lewy (CFL) criterion, which states that the time step must be sufficiently small compared to the square of the space step to guarantee the stability of the explicit scheme. By choosing (Δt) and (Δx) appropriately, the finite-difference method can provide accurate and stable solutions for the Allen-Cahn equation, enabling phase separation and pattern formation dynamics to be captured efficiently. Figures (12) and (13) illustrate examples of the calculated concentration and patterns formed using this method, highlighting the importance of discretization parameters in the accuracy of the results obtained.

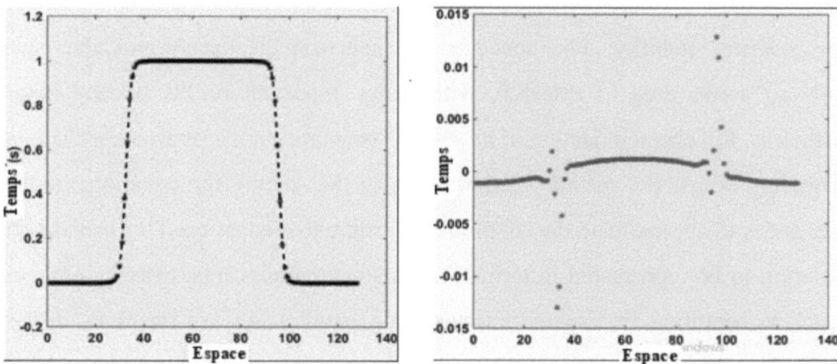

Fig. 12. Concentration with (MDF)

Fig. 13. Concentration using the spectral method

The Allen-Cahn equation is a reaction-diffusion equation that governs the dynamics of phase separation in binary alloys, where the concentration of a component varies in different regions. This equation incorporates a diffusion term and a reaction term, which facilitate phase separation by pushing the system towards a lower-energy state. To analyze the system, we employ both spectral and finite-difference methods. These methods enable us to calculate the concentration profile, follow its evolution over time, and determine the interfacial energy using both techniques. The finite-difference method (FDM) for calculating concentration is a widely used numerical approach to solving the Allen-Cahn equation. It has been successfully applied to a variety of physical and engineering problems involving phase separation and patterning (Fig. 12). The stability and accuracy of the finite-difference method for calculating the concentration in the Allen-Cahn equation depend on the selection of time and space steps (Fig. 13). The concentration spectrum of the Allen-Cahn equation is a subject of ongoing research in mathematical physics and nonlinear analysis, leading to several significant discoveries. However, a complete understanding of the spectral properties of the equation remains an open problem. The concentration spectrum is closely linked to the stability of stationary states in the Allen-Cahn equation. More precisely, the concentration values of the differential operator in the equation determine the exponential growth or decay rates of perturbations around stationary states. The presence of positive concentration values indicates instability of the stationary state, while the absence of positive concentration values indicates stability. The concentration spectrum of the Allen-Cahn equation remains an active area of research, with many important results already obtained. Nevertheless, full characterization of its spectral properties remains an open challenge.

However, we use the spectral method to solve the Allen-Cahn equation, employing Fourier series to approximate the solution of a differential equation. This method allows the solution to be represented in terms of frequency components, offering an accurate approach to capturing fine spatial variations. In parallel, we use the finite difference method, a numerical technique that approximates the derivatives of a function by finite differences calculated between discrete points. This method is particularly useful for solving differential equations on a discrete grid, facilitating the implementation of temporal and spatial discretization schemes. To evaluate interfacial energy, we apply two distinct methods: the first is based on the concentration profile and consists in integrating the interfacial energy density across the interface; the second method is based on the gradient of the concentration profile, directly calculating the energy associated with spatial variations in concentration. Interfacial energy is an essential measure of the energy required to create an interface between two distinct phases. It plays a crucial role in the study of phase separation, as it influences the dynamics and stability of interfaces. Next, we plot the difference between the concentration profiles obtained with the two methods, enabling potential discrepancies to be visualized. This comparison is fundamental to assessing the accuracy and reliability of the numerical solutions derived from each method. We also compare the interfacial energy results obtained by the two approaches in a table (Table 1).

Column labels	Finite Difference Energy	Spectral Energy
Density	0.16215	0.16643
Gradient	0.15999	0.16698
Total	0.32215	0.33341

Table. 1 Interfacial energy results.

This comparison provides a valuable check on the accuracy of the numerical solutions, highlighting the strengths and limitations of each method. Ultimately, this comparative analysis provides a better understanding of the performance of spectral and finite difference methods in the context of solving the Allen-Cahn equation.

The Allen-Cahn equation can also be solved numerically using the energy finite difference method (EFDM). Specifically, the finite-difference method for concentration in the Allen-Cahn equation can be formulated as follows:

$$u_i^{(n+1)} = u_i^{(n)} + \Delta t \left[\frac{\epsilon^2 \left(u_{i+1}^{(n)} - 2u_i^{(n)} + u_{i-1}^{(n)} \right)}{(\Delta x)^2} - u_i^{(n)} \left(u_i^{(n)} - 1 \right) \left(u_i^{(n)} + 1 \right) \right] \tag{II.12}$$

Where $(u_i^{(n)})$ is the numerical approximation of the order parameter $\big(u(x_i, t_i)\big)(\Delta t)$ and (Δx) are the time and space steps, respectively, and the exponents $((n)$ and $(n+1)$ refer to time steps (t_n) and (t_{n+1}) respectively. The first term on the right-hand side of the equation represents the diffusion term, while the second term represents the nonlinear reaction term. In the finite-difference approximation (FDM), these terms are evaluated at discrete spatial and temporal grid points. The stability and accuracy of the finite-difference method for concentration in the Allen-Cahn equation depend on the choice of time and space steps. The time step must satisfy the Courant-Friedrichs-Lewy (CFL) condition, which requires that:($\Delta t \leq (\Delta x)^2/(4\epsilon^2)$). In addition, the spatial step must be small enough to resolve the transition layer between the two phases of the binary mixture. On the other hand, the energetic finite-difference method (EFDM) is an approach that combines the finite-difference method with energetic principles to solve partial differential equations, including the Allen-Cahn equation. In the EFDM approach applied to the Allen-Cahn equation, the order parameter u(x, t) is treated as an energy potential function, and the numerical approximation of (u) is obtained by minimizing the total potential energy of the system. The total potential energy is defined as the sum of the potential energy contributions of each discretized point in the spatial domain. EFDM discretizes the spatial domain into a finite set of discrete points and approximates the order parameter as a piecewise linear function defined on the grid points. The temporal evolution of the order parameter is determined by minimizing the total potential energy of the system while preserving physical properties such as conservation of mass and energy. EFDM for the Allen-Cahn equation offers several advantages over traditional finite-difference methods. For example, it naturally conserves mass and energy, and is free from spurious oscillations and other numerical artifacts. However, EFDM is more computationally expensive than traditional finite difference methods, and may require specialized algorithms and software. In the ESM (spectral energy method) approach to the Allen-Cahn equation, the order parameter $(u(x,t))$ is represented as a Fourier series:

$$u(x,t) = \sum_k 1^N [\, ak(t)\cos(kx) + bk(t)\sin(bx)] \qquad \text{(II.13)}$$

Where (ak) and (bk) are the Fourier series coefficients, and N is the number of Fourier modes used in the approximation. The time evolution of coefficients (ak) and (bk) is determined by solving a system of ordinary differential equations, derived by applying the Fourier transform to the Allen-Cahn equation. This approach enables us to accurately capture high-frequency oscillations and other rapid changes present in the solution. We can obtain an accurate representation of high-frequency oscillations and other rapid changes in the solution using EFDM. What's more, EFDM is computationally efficient, especially for problems with periodic boundary conditions. However, the Energy Spectral

Method (ESM) has certain limitations. For example, it may not be suitable for problems involving irregular boundaries or complex geometries, and it may require the use of specialized algorithms and software. Nevertheless, the energy spectral method remains a powerful numerical approach capable of accurately and efficiently solving the Allen-Cahn equation, as well as other partial differential equations.

Solving the Allen-Cahn equation coupled to the Navier-Stokes equations using finite differences involves a combination of finite difference methods for both equations. Here is a basic code illustrating the numerical solution. This study presents a numerical approach to simulating the dynamics of a three-dimensional droplet undergoing phase separation in a fluid governed by the Allen-Cahn equation coupled to the Navier-Stokes equations. The numerical method used employs finite-difference techniques to discretize the partial differential equations, enabling the evolution of droplet morphology and fluid flow characteristics to be explored (Fig 14. a. b. c. d). The Allen-Cahn equation is discretized in time using an explicit finite-difference scheme. Spatial derivatives are approximated by central differences. A time step is taken to update the order parameter (ϕ). The discretization of the incompressible Navier-Stokes equations is performed using a shifted grid approach. In addition, a two-stage method is implemented to partition the solution into prediction and correction stages. Then, an iterative process is used to update the velocity and pressure fields, guaranteeing the satisfaction of continuity constraints. The Allen-Cahn equation is coupled to the Navier-Stokes equations through the advection term in the Allen-Cahn equation and the volume force term in the Navier-Stokes equations. An iterative coupling strategy ensures consistency between the order parameter and the fluid flow fields.

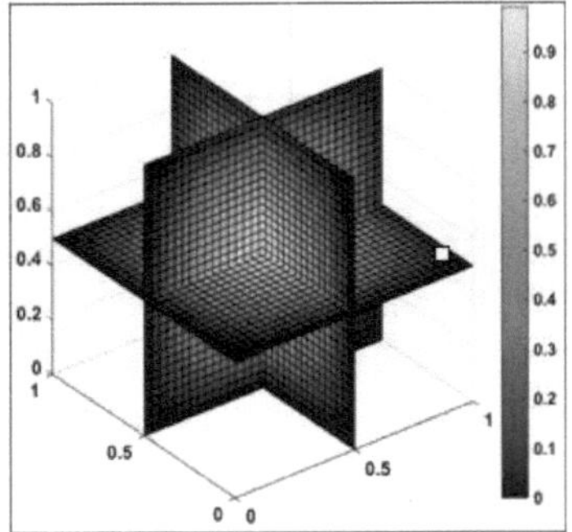

Fig. 14.a. Solving the Allen-Cahn / Navier-Stokes equations in 3D (0 seconds).

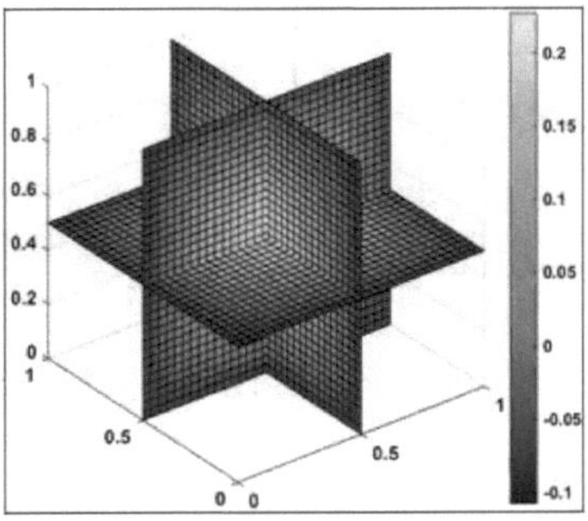

Fig. 14.b. Solving the Allen-Cahn / Navier-Stokes equations in 3D (0.2 seconds).

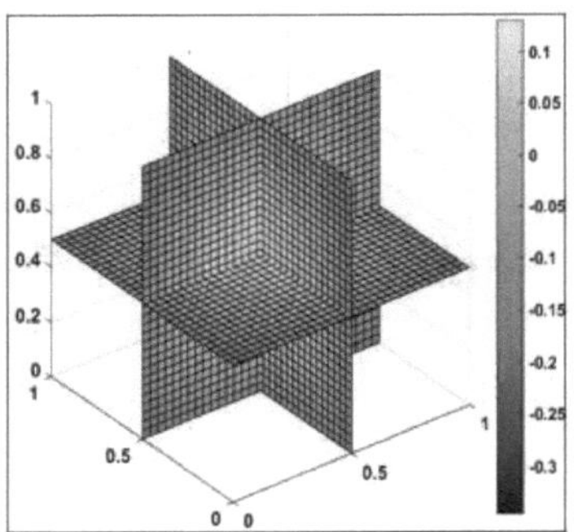

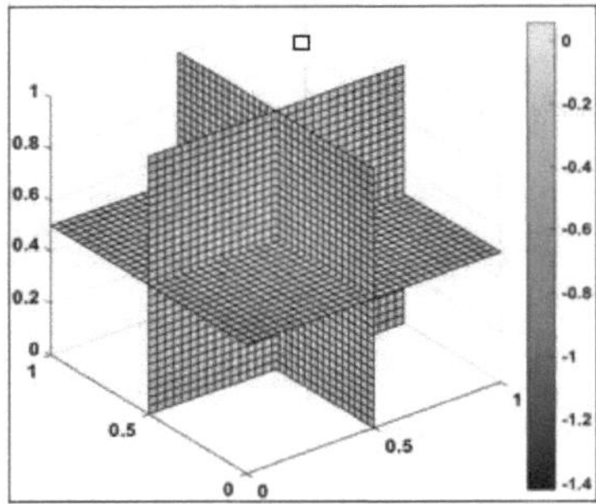

Fig. 14.c. Solving the Allen-Cahn / Navier-Stokes equations in 3D (0.3 seconds).

Fig. 14.d. Solving the Allen-Cahn / Navier-Stokes equations in 3D (0.5 seconds).

The Allen-Cahn equation and the Navier-Stokes equations are interconnected by incorporating the advection term of the Allen-Cahn equation and the force-density term of the Navier-Stokes equations. To ensure consistency between the order parameter and the fluid flow fields, an iterative coupling strategy is employed. The simulation effectively illustrates the dynamic evolution of droplet morphology, highlighting phase separation phenomena over time. Visualization techniques, including renderings (3D) and cross-sections, are used to analyze the internal structure of the droplet. The numerical approach skilfully models the complex interplay between phase separation and fluid flow in three dimensions. In addition, the numerical methodology presented establishes a solid framework for the study of the Allen-Cahn and Navier-Stokes equations in (3D). The insights gained from these simulations contribute significantly to advancing our understanding of droplet dynamics in complex fluid environments.

Conclusion

In this study, we presented a numerical iterative method for approximating the solutions of the nonlinear Cahn-Hilliard-Navier-Stokes equations. The numerical tests carried out demonstrate the accuracy and efficiency of this approach, offering robust approximations to the solutions of these complex equations. By comparing our results with other experimental and numerical approaches, we have validated the superiority of our method in terms of simulation performance and fidelity. These results underline the importance of our approach for solving multi-phase flow and phase separation problems, opening up new prospects for applications in various scientific and industrial fields.

Chapter III: Experimental results

I. Introduction

The history of experimental results validating the Allen-Cahn equation goes back several decades, marking a crucial milestone in the confirmation and application of this equation in physics and materials science. The first attempts at experimental validation of the Allen-Cahn equation date back to pioneering work in the field of phase separation in binary alloys. Over time, researchers have carried out a series of experiments to compare theoretical predictions of the Allen-Cahn equation with empirical observations of actual phase separation in various materials. These experiments often involved the use of advanced microscopic imaging and analysis techniques to study phase morphology, phase separation kinetics, as well as the measurement of concentration profiles. Experimental results have helped to refine and validate physical parameters and coefficients of the Allen-Cahn equation, such as interfacial energy, diffusion, and other parameters of interface kinetics. This experimental validation is essential to assess the ability of the Allen-Cahn equation to accurately model complex phenomena such as patterning, coalescence and phase growth in real physical systems. As a result, experimental results have reinforced the relevance and applicability of the Allen-Cahn equation in a variety of scientific and technological fields, from metallurgy and polymer materials to materials physics and biology. They continue to play a crucial role in the improvement of theoretical models and in technological innovation based on the understanding of phase separation phenomena.

II. Results

In order to precisely define the error and asymptotic bounds, extensive research was carried out on the numerical results. This research covered a wide range of approaches, from numerical experiments to fine analysis of behavior near singularities. The Allen-Cahn equation, recognized for its conservative nature, is widely used as a model to study the dynamics of moving interfaces between two phases. These phases can include interfaces such as liquid-gas or solid-liquid. This mathematical model makes it possible to accurately capture and predict how these interfaces evolve over time, in response to gradients in temperature, concentration or other factors influencing phase transitions. This detailed version clarifies how research has addressed numerical and theoretical aspects to delineate errors and asymptotic behaviors, using the Allen-Cahn equation as the main tool for modeling phase interfaces in different systems. This equation takes into account the influence of fluid velocity on the interface (u).

$$\frac{\partial \phi}{\partial t} + \nabla.(\phi u) = \nabla.[M(\nabla \phi - \xi n)] \tag{III.1}$$

Where, parameter (λ), on the right-hand side of the Allen-Cahn equation, is the norm of the gradient vector of (ϕ_{eq}) at equilibrium. To improve the quality of the simulations and refine their accuracy, the droplet distribution, characterized by a Rosin-Rammler distribution (also known as the Rosin-Rammler-Sperling-Bennett or RRSB distribution), is used in the model. Two distinct representations of the experimental droplet distribution are employed in the simulations. This subsection focuses on explaining the Rosin-Rammler representation, offering an in-depth exploration of the underlying theory governing this distribution.

$$F(D) = e^{-\left(\frac{D}{d}\right)^{n}} \tag{III.2}$$

This is a mathematical model that describes the cumulative distribution function (CDF) of a population of particles or droplets as a function of their size. In this equation, ($D=60\mu m$) represents the droplet diameter, (d) is the characteristic diameter, and ($n=2.25$) is the distribution parameter. The parameters (d) and (n), obtained from the fit, are used to describe the droplet size distribution in your system. For validation purposes, we systematically compare our numerical results with established data documented in existing scientific literature. This benchmarking is a crucial step in assessing the reliability and accuracy of our simulation results in relation to previously validated results. By aligning our simulation conditions with those detailed in the literature, we aim to validate the fidelity of our model and ensure its ability to accurately represent the physical phenomena under investigation (Fig. 15 and Fig.16). Our numerical simulations show notable agreement with the experimental results, underlining the robustness and accuracy of our numerical model in faithfully representing the physical phenomena observed. This convergence between our simulated and experimental data constitutes a convincing validation of our simulation methodology, reinforcing the credibility of our research results.

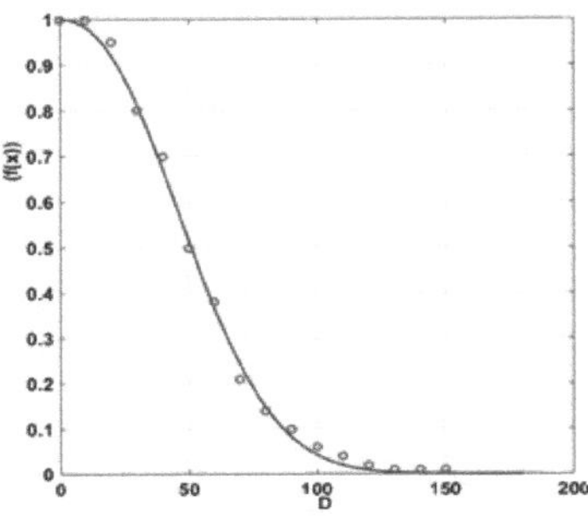

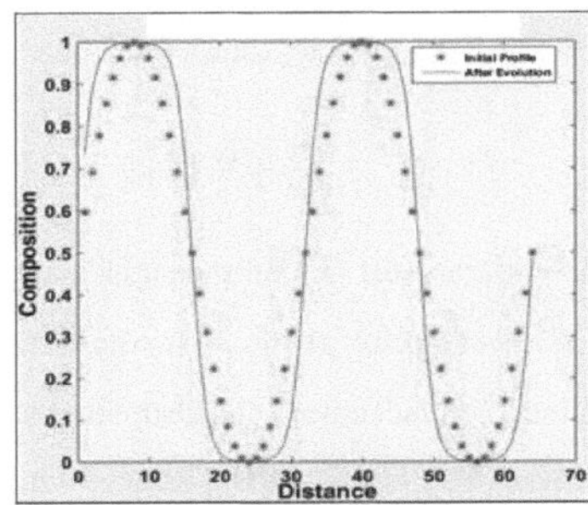

Fig. 15. Rosin-Rammler droplet size distribution.

Fig. 16. Allen-Cahn equation profile for two immiscible, incompressible fluids.

As distance increases, the droplet profile tends to approximate the compositional value observed in the bulk region of the fluid. When a droplet is placed in a simple shear flow, it undergoes deformation and stretching due to the velocity gradient present in the flow (Fig. 17). The temporal dynamics of the droplet are influenced by several factors, such as the viscosity ratio between the droplet and the surrounding fluid, the initial shape of the droplet and the flow conditions. Generally speaking, under the effect of shear forces, the droplet adopts an ellipsoidal shape. The rate of deformation depends on the intensity of the flow, the size of the droplet and the viscosity ratio between the droplet and the surrounding fluid. When the droplet is initially spherical, it stretches in the direction of the shear flow, modifying its cross-section to become elliptical. As deformation continues, the droplet may manifest various instabilities, such as dripping or rupture. The precise behavior of the droplet depends on the flow parameters and intrinsic characteristics of the droplet, as well as on the properties of the fluid interfaces. Several theoretical and numerical models have been developed to study the temporal evolution of droplets in simple shear flows, notably through numerical simulations using the Boltzmann lattice method, the immersed boundary method and the boundary element method. In addition, experimental techniques such as micro-fluidic devices and high-speed imaging are also employed to observe and measure droplet deformation and instability in shear flows. Overall, the study of the temporal evolution of droplets in simple shear flows remains a complex and active area of research in the field of fluid mechanics.

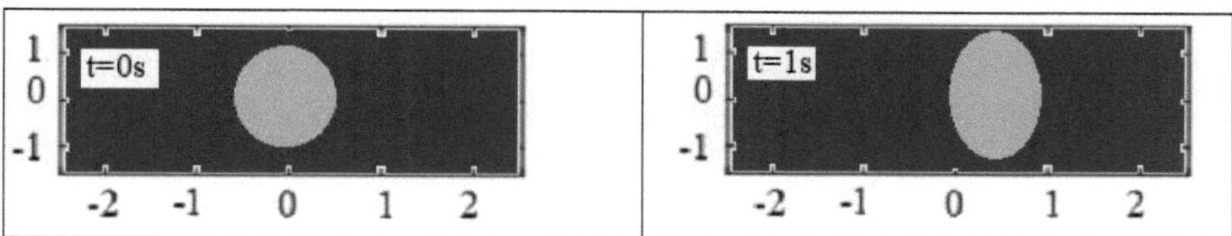

Fig. 17. Time evolution of a droplet in shear flow.

The mathematical model for predicting bubble size and frequency using the Allen-Cahn-Navier-Stokes model involves a set of partial differential equations that describe the evolution of fluid flow and phase field variables. Firstly, the conservative Allen-Cahn equation represents a particular form of the Allen-Cahn equation, a partial differential equation (PDE) used to describe how phase separation evolves in a binary system. This equation models how the concentration of a component varies across different regions of the system over time, in response to diffusion and reaction forces. The continuity equation ensures the incompressibility of the fluid and is given by :

$$\left(\frac{\partial u}{\partial t} + \nabla.(\boldsymbol{u} \otimes \boldsymbol{u})\right) + \boldsymbol{\nabla p} = \nabla.(2\mu\boldsymbol{\nabla u} - \xi(\nabla.\boldsymbol{u})\boldsymbol{I}) + \boldsymbol{f} \qquad \text{(III.3)}$$

ξ is the volume viscosity. The conservative form of the Allen-Cahn equation explicitly includes a convective term $(\nabla.(\boldsymbol{u} \otimes \boldsymbol{u}))$which takes into account the advection of the order parameter $(\boldsymbol{u})$.

The equation is often coupled to the incompressible Navier-Stokes equations to model the simultaneous evolution of fluid flow and phase separation. Information on bubble size and frequency can be extracted from simulation results of the phase field variable $(\boldsymbol{u})$.

The regions where ($\boldsymbol{u}$) transitions between two distinct values correspond to the presence of bubbles. To create a graphical representation of the final configuration of a deformed droplet in the central xy-plane, it is necessary to define the shape or distortion of the droplet. Starting from specific initial conditions (distortion amplitude L = 1 at x = 0, y = 0), we illustrate that the droplet is deformed by a sinusoidal distortion (Fig. 18.a). In fluid dynamics, the vortex measures the local rotation of fluid elements. In a two-dimensional flow, the vortex vector has only one non-zero component, often represented as the scalar vortex. The vortex (ω) is defined as the loop of the velocity vector ($\omega = \nabla \times u$). In 2D flow, the velocity vector can be written as ($u = u_1, u_2, 0$)where (u_1) is the velocity component in the (x) direction, (u_2) is the velocity component in the x direction. y direction, and there is no velocity component in the (z) direction (Fig. 18.b).

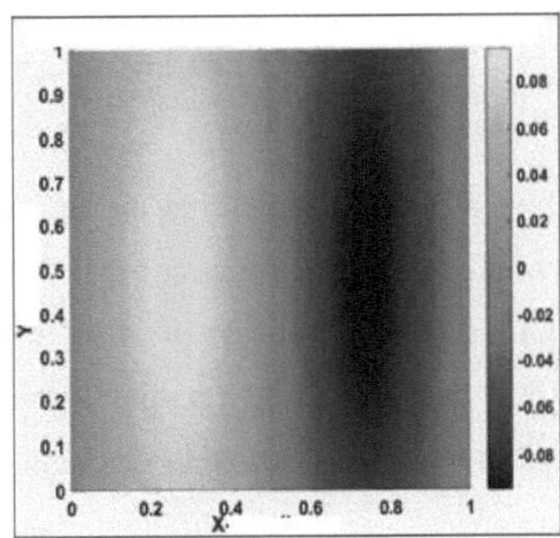

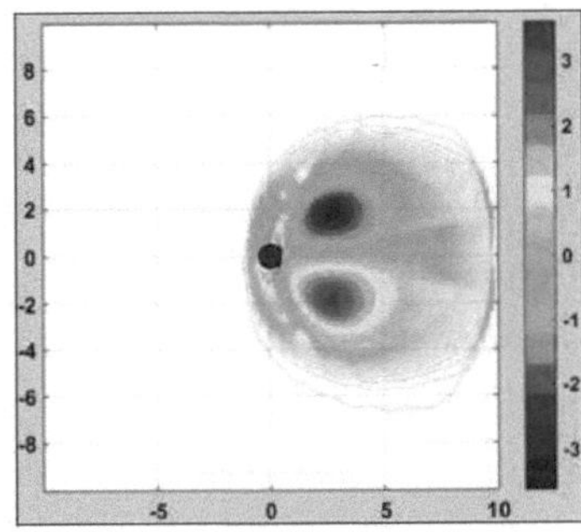

Fig. 18.a. Final configuration of the deformed droplet (x-y plane).

Fig. 18.b. 2D droplet verticality components (x-y).

Water droplet impact force refers to the amount of force applied to a droplet when it comes into contact with a surface or other object. The effects of this force depend on a number of factors. First of all, the size of the droplet plays a crucial role in the magnitude of the impact force felt. A larger droplet will tend to exert a more significant force on collision than a smaller droplet. The speed at which the droplet hits the surface is another determining parameter. A higher velocity increases the droplet's kinetic energy, thus increasing the perceived impact. Conversely, a lower speed can generate a more moderate impact force. Finally, the nature of the surface or object with which the droplet comes into contact is crucial. A rigid, smooth surface may reflect part of the impact force, while an absorbent or irregular surface may absorb part of this force, thus affecting the results of the interaction. In short, the force of impact on water droplets varies according to droplet size, velocity at impact, and the composition and texture of the surface or object in contact.

$$F_D = \frac{mv^2}{\text{Diameter}} \tag{III.4}$$

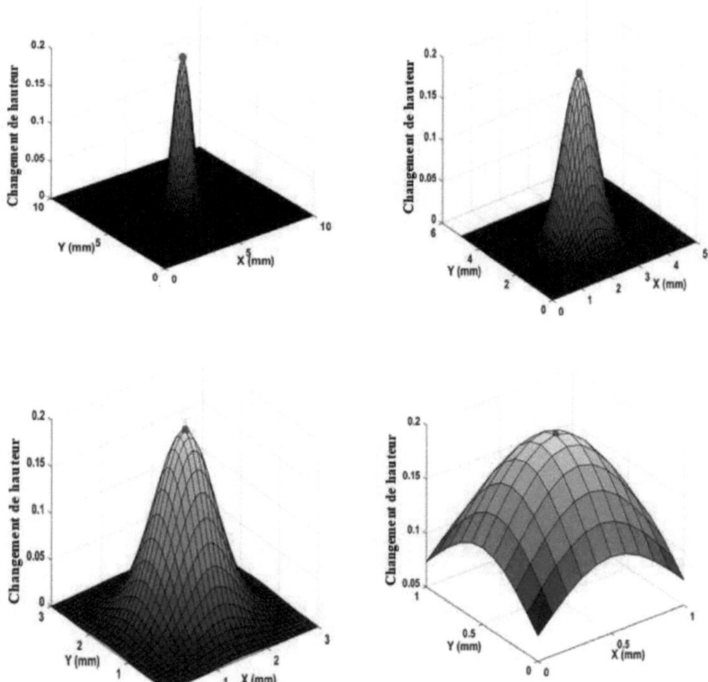

Fig 19. Simulated droplet impact morphology.

When a water droplet strikes the surface of a face mask, several stages characterize its interaction. Firstly, at the moment of initial contact, the droplet begins to deform under the effect of the rapid deceleration resulting from the transfer of its kinetic energy to the surface. This rapid deceleration results in a rapid increase in impact force, while the droplet's velocity gradually decreases. As the droplet continues to deform on impact, it eventually reaches a point of maximum compression, where it is most compressed. Depending on the initial energy of the droplet and the specific properties of the mask surface, it may either rebound after this phase, or remain in contact with the surface. During this phase of maximum compression, the impact force can reach its peak, mainly due to the rapid change in the droplet's momentum. This dynamic is illustrated in (Fig. 19).

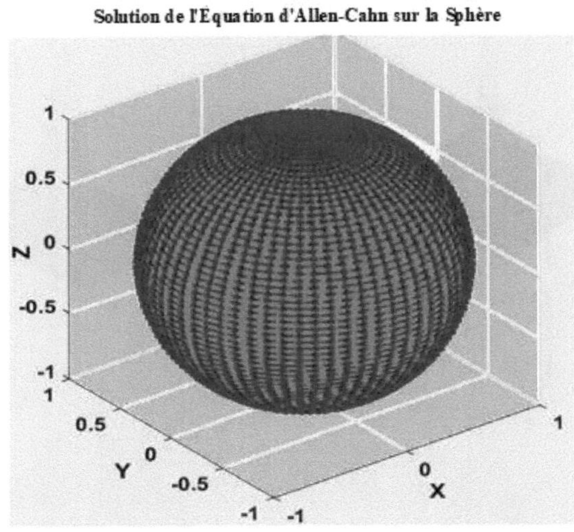

Fig. 20. Evolution of the numerical solution of the Allen-Cahn equation on the sphere.

Another method is the finite element method, which discretizes the sphere into smaller elements and solves the equation numerically by approximating the solution in each element. This method is more flexible and can handle higher dimensions, but requires more computational resources. Over the years, various improvements have been made to these numerical methods to enhance their accuracy and efficiency. For example, adaptive mesh refinement techniques can be used in the finite element method to refine the mesh in regions where the solution exhibits rapid variations, while coarsening the mesh in regions with slow variations. This approach can considerably reduce computational cost while preserving accuracy. In recent years, deep learning methods have also been used to solve the Allen-Cahn equation on the sphere. One approach is to use neural networks to approximate the solution directly, thus eliminating the need for discretization. Another approach is to use neural networks to speed up existing numerical methods by learning the solution at specific points and interpolating it elsewhere. Overall, the evolution of numerical solutions for the Allen-Cahn equation on the sphere has been driven by a combination of mathematical and computational advances, leading to increasingly accurate and efficient solution methods for this significant equation.

In this section, we have drawn on existing experimental data in the literature. This study focuses on investigating the dynamics involved in the impact of water droplets at low velocities. This research explores the complex processes that occur when water droplets collide with surfaces under low velocity conditions. The aim is to gain a comprehensive understanding of the behavior of droplets on impact, including their deformation, spreading and interaction with the face mask surface. Through detailed experiments and analysis, this study aims to contribute to a better understanding of the mechanics of low-velocity water droplet impact. The results obtained can potentially lead to advances in several fields, such as surface coatings, spray technologies and natural processes involving droplet interactions, such as raindrop impacts on face mask surfaces. By better understanding these mechanisms, it becomes possible to improve surface designs and humidity control technologies, as well as optimize practical applications in various industrial and environmental sectors. We base our study on the transient force curve of a droplet impacting the sensor at a velocity of 3.27 (m/s). The raw data contain significant oscillations, which can complicate accurate analysis of impact dynamics. To attenuate these oscillations, all impact force data were filtered using a low-pass finite impulse response (FIR) filter with a cut-off frequency of 10,000 Hz. This filtering technique successfully eliminated unwanted oscillations, making the data smoother and easier to interpret. However, the filtered curve was shifted by around 100 μs compared with the original experimental data. Despite this time shift, the essential features of the original force data, such as the rapid rise time and prolonged fall time, were preserved.

These features are crucial for understanding force behavior at the moment of impact. Consequently, the filtered curve has been used as a general approximation to describe the evolution of force during the impact of a water droplet. This approach provides a more faithful and usable representation of the impact dynamics, facilitating subsequent analyses and comparisons with other studies. Using this method, we follow the work of Mitchell et al. (2016), who demonstrated the effectiveness of this type of filtering for droplet impact analysis (Fig. 21).

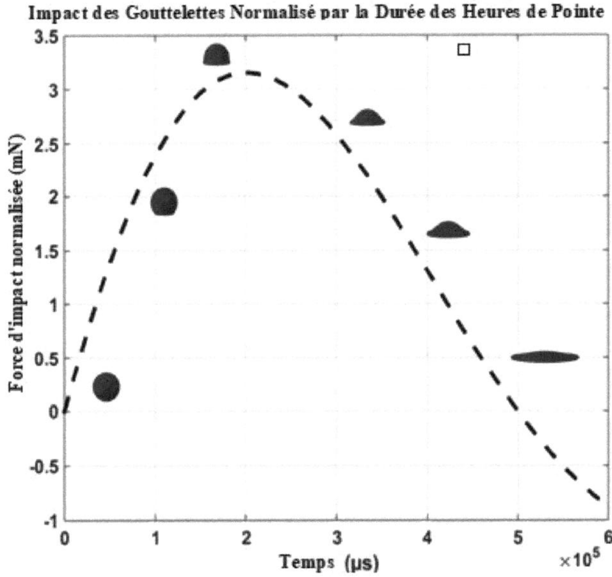

Fig. 21. Comparison between numerical simulations (impact force prediction) and experimental results of droplet deformation (Mitchell et al, 2016).

In this experiment, the researchers used a Model 482 PCB signal conditioner to process the signals from the sensors. This signal conditioner is essential for amplifying, filtering and converting raw analog signals into a more usable and accurate form, thus facilitating the analysis of experimental data. In addition, a Le Croy Wave Surfer 64MXs-B oscilloscope was used to record the measurements. This oscilloscope, renowned for its high performance and accuracy, was set to sample at a frequency of 20 MHz for all experimental measurements. This high sampling rate is crucial for capturing the fast dynamics and fine details of the forces involved in droplet impact. The combination of these instruments enables reliable and accurate data acquisition. The signal conditioner ensures that signals are properly prepared before recording, while the oscilloscope provides sufficient temporal resolution to observe rapid variations in forces. Together, this equipment has enabled a detailed and accurate analysis of the phenomena observed during droplet impact experiments, as illustrated in (Fig. 21).

General conclusion

The Cahn-Hilliard equation has been widely used to study phase separation in various systems, including polymer blends, spinodal decomposition in alloys, and refining in foams and emulsions. In polymer blends, the Cahn-Hilliard equation describes how the different components separate and form distinct domains over time. It can be used to predict the formation of complex structures, such as alternating layers or interpenetrating networks, resulting from the interplay between diffusion and surface tension. The equation is also used to model spinodal decomposition, a process by which a homogeneous alloy breaks down into two distinct phases due to concentration fluctuations. It helps to understand how microstructures evolve and refine over time, influencing the mechanical and physical properties of materials. In foams and emulsions, the Cahn-Hilliard equation is used to model the stability of interfaces between phases and the coalescence of droplets or bubbles. It is used to optimize material properties by adjusting process parameters to obtain structures with specific characteristics. The equation describes how molecules move and redistribute to minimize the free energy of the system. It incorporates the effects of surface tension, which tends to minimize the interface area between phases. Also, the Cahn-Hilliard equation is based on a free energy potential that determines phase stability and pattern formation. By enabling the prediction and control of material microstructure, the Cahn-Hilliard equation is essential for the development of materials with optimized properties. Its applications extend to many industrial sectors, including aerospace, automotive, medical technologies and electronic devices, where material performance is critical. Overall, the Cahn-Hilliard equation is a powerful tool for understanding the dynamics of phase separation and multi-phase flows. Its ability to capture the development of patterns and structures makes it a significant tool for the design and optimization of materials and processes. It continues to play a key role in scientific research and technological development, offering prospects for new innovations and improvements in a variety of application fields.

The Cahn-Hilliard equation plays a crucial role in the study of multi-phase flows in fluid dynamics. This equation is used to model and analyze complex phenomena such as droplet formation, coalescence and breakup. Here's a detailed look at its application and coupling with other equations to understand and simulate these phenomena. The Cahn-Hilliard equation simulates how droplets form from a continuous phase, taking into account interfacial forces and diffusion. It helps to understand the factors influencing the size and shape of newly formed droplets. The equation models the process by which two or more droplets merge to form a larger droplet, by describing the dynamics of the interface at the molecular level. Variables such as surface tension and phase viscosity affect the rate and nature of coalescence, captured by the equation. The Cahn-Hilliard equation describes how droplets can fragment under external forces, such as velocity gradients or turbulence. Droplet stability is evaluated as a function of the free energy of the system, influencing the conditions of rupture.

To model complex multi-phase flows, the Cahn-Hilliard equation is often coupled with the Navier-Stokes equations. This coupling takes into account not only interface dynamics, but also interactions with the velocity field of the surrounding fluid. In micro-fluidic devices, this coupling is used to simulate fluid behavior at very small scales, where phase interactions are critical. The model is also applicable to study flows in porous media, where interfaces between phases can move through complex structures. In the chemical and petrochemical industries, the model helps optimize phase separation processes, improving efficiency and product quality. The model contributes to the design of reactors where multi-phase flows are common, enabling better understanding and control of reactions. In fluid dynamics, the Cahn-Hilliard equation is a powerful tool for studying and modeling multi-phase flows, including droplet formation, coalescence and break-up. Its coupling with the Navier-Stokes equations makes it possible to simulate complex systems, such as those encountered in micro-fluidics, porous media and various industrial processes. This combination offers a robust approach to understanding interface dynamics and enhancing practical applications in many areas of science and engineering. The study of two immiscible and incompressible fluids using the coupled Allen-Cahn and Navier-Stokes equations represents an important area of research in fluid mechanics. The Allen-Cahn equation governs the evolution of the interface between the fluids, while the Navier-Stokes equations describe fluid motion and flow. The coupling of these two equations provides a complete model of the behavior of two immiscible fluids. The coupled Allen-Cahn and Navier-Stokes equations can be solved numerically using a variety of methods, including finite difference, finite element or spectral methods. These methods involve discretizing the equations in space and time, and solving the resulting algebraic equations using iterative methods. The study of immiscible and incompressible fluids using the coupled Allen-Cahn and Navier-Stokes equations is essential for understanding the complex phenomena of phase separation and fluid flow. The various numerical methods available offer powerful tools for simulating these phenomena with increased accuracy and efficiency, enabling significant advances in research and industrial applications.

Numerical results of concentration evolution obtained from the Allen-Cahn equation provide valuable information on phase separation and pattern formation. These results are influenced by factors such as initial conditions, equation parameters and boundary conditions. The evolution of the concentration of a binary mixture using the Allen-Cahn equation can thus contribute to optimizing the design and performance of materials and processes. The numerical study of the Allen-Cahn equation offers a powerful tool for understanding and predicting phase separation and patterning in binary systems. Simulations explore the influence of initial conditions, equation parameters and boundary conditions on concentration evolution. The results obtained are essential for optimizing the design and performance of materials, paving the way for innovations in various fields of materials science and engineering. In fluid simulation, iso-surfaces are used as visual representations of the interface between two fluids. They enable the study of dynamic phenomena such as vortex behavior and wave formation at the interface, corresponding to a specific value of a scalar field. Interfacial tension influences fluid dynamics by introducing a force term into the Navier-Stokes equations, which takes into account the impact of interfacial tension on fluid velocity and pressure. These numerical simulations validate the effectiveness and reliability of the NS-AC model in accurately representing the complex dynamics of multi-fluid flows. The model has proved both accurate and efficient compared with other numerical methods, making it applicable to a variety of scientific and engineering problems.

General Conclusion

The Cahn-Hilliard equation finds wide application in studying phase separation, whether in polymer blends, spinodal decomposition in alloys, or refinement in foams and emulsions. In polymer blends, it predicts the formation of complex structures like alternating layers or interpenetrating networks, influenced by diffusion and surface tension. It also models spinodal decomposition, where a homogeneous alloy separates into two distinct phases due to concentration fluctuations, affecting mechanical and physical properties of materials. In foams and emulsions, it optimizes interface stability and the coalescence of droplets or bubbles by adjusting parameters to achieve specific structures. By integrating the effects of surface tension and based on a free energy potential, the Cahn-Hilliard equation enables prediction and control of microstructure, playing a crucial role in developing materials with optimized properties. Its applications span diverse sectors such as aerospace, automotive, medical technologies, and electronic devices, contributing to understanding and improving material processes and performances.

In fluid dynamics, the Cahn-Hilliard equation is essential for modeling complex multiphase flows. It describes the formation, coalescence, and rupture of droplets, accounting for interfacial forces and diffusion. Coupled with Navier-Stokes equations, it simulates these phenomena across various scales, from microfluidic devices to industrial porous media. This coupling provides a robust approach to understand and predict interface dynamics, enhancing practical applications across scientific and engineering domains.

Outlook

The prospects for this research work are vast and promising, offering several directions for future developments and potential applications:

1. **Numerical Model Optimization**: Continue to improve the numerical methods used to solve the Allen-Cahn equation and the coupled Navier-Stokes equations. This could include the development of more efficient numerical solution techniques, such as adaptive methods or domain decomposition methods, to increase accuracy and reduce computational cost.
2. **Study of more complex cases**: Apply the NS-AC model to more complex cases of multiphase fluids, for example by incorporating more realistic boundary conditions or studying more complex fluid geometries and configurations. This would enable the model to be validated in a wide range of real-life situations.
3. **Experimental validation**: Collaborate with experimentalists to validate numerical results with real experimental data. This would confirm the accuracy of the model in practical scenarios and identify the key parameters needed for a good match with experimental observations.
4. **Industrial applications**: Explore the potential applications of the NS-AC model in industry, for example to optimize mixing processes, reactor design, or the simulation of fluids in micro-fluidic devices. This could open up opportunities to solve complex engineering problems and improve the efficiency of industrial processes.
5. **Developing New Capabilities**: Extend the model's capabilities to include additional physical effects such as thermo-capillarity, electromagnetic interactions, or other phenomena specific to particular applications. This could make the model even more versatile and adaptable to an even wider range of problems.

In summary, this research work offers many opportunities to extend and apply the NS-AC model in a variety of scientific and technological contexts, helping to advance our understanding of multiphase flows and improve practical solutions in many industrial and engineering fields.

Bibliography

1. Y. Hairch , A. Elmelouky, M. Louzazni, F. Belhora and M.Monkade, A numerical study of interface dynamics in fluid materials. *Journal of Matériaux & Techniques* (2024).
2. Y. Hairch, A. Jraifi, I. Medarhri, A. Elmelouky, Mathematical Modeling of Mechanical Properties in the Permeation of Green Hydrogen through Membrane Separation Materials. *Journal of mathematical modeling and computing* (2024).
3. Y. Hairch, R. Elotmani, A. Elmelouky, M. Mansouri, Exploring the Mechanical Dynamics and Physical Characteristics of Droplets Using Face Mask Materials. *Euro-Mediterranean Journal for Environmental Integration* (2024).
4. Y. Hairch, R. Mghaiouini, A. Mortadi, D. Saifaoui, M. Salah, A. Graich, E. Chahid, A. Elmelouky, M. Monkade and A. El Bouari, Modeling and simulations of moving droplets in relation to SARS-CoV-19 generated by respiratory system. *Journal of Aerosol Science and Engineering* (2022).
5. Hairch, Y., El Afif, A., 2020. Mesoscopic modeling of mass transport in viscoelastic phase-separated polymeric membranes embedding complex deformable interfaces, *J Membrane Science, Elsevier* 596, 117589 (2020).
6. El-Atab, N., Qaiser, N., Badghaish, H.S., Shaikh, SF., Hussain, MM., 2020. Flexible Nanoporous Template for the Design and Development of Reusable Anti-COVID-19 Hydrophobic Face Masks. *J ACS Nano* (6), 7659-7665 (2020).
7. Fadare, OO, Okoffo, ED., 2020. Covid-19 face masks: A potential source of microplastic fibers in the environment. *J Science of the Total Environment*. Elsevier (737), 140279 (2020).
8. Zhou, L., Nyberg, K., Rowat, AC., 2015. Understanding diffusion theory and Fick's law through food and cooking. *J Advances in Physiology Education* 39(3), 192-197 (2015).
9. Luzi, F., Torre, L., Kenny, J., Puglia, D., 2019. Bio- and Fossil-Based Polymeric Blends and Nanocomposites for Packaging: Structure-Property Relationship". *J Materials* 12(3), 471 (2019).

10. Malik, T., Razzaq, H., Razzaque, S., Nawaz, H., Siddiqa, A., Siddiq, M., Qaisar, S., 2018. Design and synthesis of polymeric membranes using water-soluble pore formers: an overview. *J Polymer Bulletin*, Springer 76, 4879-4901 (2018).
11. Zhu, J., Hou, J., Zhang, Y., Tian, M., Tao, H., Liu, J., Chen, V., 2017. Polymeric Antimicrobial Membranes Enabled by Nanomaterials for Water Treatment. *J Membrane Science, Springer* 17, S0376-7388 (2017).
12. Mirikar, D., Palanivel, S., Arumurua, V., 2021. Droplet fate, efficacy of face mask, and transmission of virus-laden droplets inside a conference room. *J Physics of Fluids* 33, 065108 (2021).
13. Rahman, M. Z., Hoque, M. E., Alam, M. R., Rouf, M.A., Khan, S. I., Xu, H., Ramakrishna, S., 2022. Face Masks to Combat Coronavirus (COVID-19)-Processing, Roles, Requirements, Efficacy, Risk and Sustainability. *Polymers journal* 14(7): 1296 (2022).
14. Daniel, B.O., Catherine, A.P., 2020. Think-Pair-Listen in the Online COVID-19 Classroom, Environmental engineering science, 3.Arup, K., S. COVID-19 and Unsafe Water: A Tale of Two Enemies. *Environmental engineering science* 37, (2020).
15. Amrit, K., Kiranmay, S., Abhinav, P., Rajeev, K M., Panuganti, C.S. D., 2022. COVID-19 Lock-down in Delhi: Uderstanding Trends of Particulate Matter in Context of Land-Use Patterns, GIS Mapping and Meteorological Traits. *Environmental engineering science* (2022).
16. Jens, I., Mikael, M.L., 2021. Multifidelity Response Surface Approximations for the Optimum Design of Diffuser Flows, *J Optimization and Engineering, Springer*. 2, 453-468 (2021).
17. Michael, H., Christian, K., Tobias, K. A., 2018. Goal-oriented dual-weighted adaptive finite element approach for the optimal control of a nonsmooth cahn-Hilliard-Navier-Stokes system. *J Optimization and Engineering*, Springer 19, pp. 629-662 (2018).
18. Zhdanov, VP, Kasemo, B., Virions and respiratory droplets in air: Diffusion, drift, and contact with the epithelium. *J BioSystems, Elsevier* 198, 104241 (2020).

19. Ataei-Pirkooh, A., Alavi, A., Kianirad, M., Bagherzadeh, K., Ghasempour, A., Pourdakan, O., Adl, R., Kiani, S.J., Mirzaei, M., Mehravi, B., 2021. Destruction mechanisms of ozone over SARS-CoV-2. *J Scientific reports* 11, 18851 (2021).
20. Dnyanesh, M., Silambarasan, P., Venugopal, A., 2021. Droplet Fate, Efficacy of Face Mask, and Transmission of Virus-Laden Droplets inside a Conference Room. *J Physics of Fluids* 32(5), 065108 (2021).
21. Scharfman, B.E., Techet, A.H., Bush, J.W.M., Bourouiba, L., 2016. Visualization of sneeze ejecta: steps of fluid fragmentation leading to respiratory droplets. *Experiments in Fluids*, Springer 57 (2016).
22. Gao, N., Niu, J., Morawska, L., 2008. Distribution of Respiratory Droplets in Enclosed Environments under Different Air Distribution Methods. *J Building Simulation, Springer* 4, 326-335 (2008).
23. Harlow, F.H., Welch, J.E., 1965. Numerical calculation of time dependent viscous incompressible flow with free surface. *Physics of Fluids* 8, 2182-2189 (1965).
24. Sleiman, GE., 2018. Processability of recycled thermoplastics. Thermorheological analysis of PP/PE blends, *thesis, University of Brittany Loire* (2018).
25. Nikolova, S., Nenov, M., 2018. Modelling vaccine quantity in mathematical models of melanoma treatment, *J Series on Biomechanics* 32, 19- 25 (2018).
26. Singh, Y., Shekhar, K., Tyagi, A. P., 2021. Mathematical modeling of mucus transport in airways due to cough: quasi - steady state turbulent condition, *J Series on Biomechanics* 35, 69-84 (2021).
27. Valentin, S., Basin, S., Chaouat, A., 2022. Severe forms of COVID-19 among patients with chronic respiratory diseases: beattentive to the severity of the underlying respiratory impairment, *J Respiratory Medicine and Research* 81, 100902 (2022).
28. Reychler, G., Vecellio, L. Dubus, J.C ., 2020. Nebulization: A potential source of SARS-CoV-2 transmission. *J Respiratory Medicine and Research, Elsevier* 78, 100778 (2020).
29. Flavio M. R. D. S. J., 2022. New Normal: The Dynamics of Air Pollutants on the Interruption-Recovery Pattern Related to the COVID-19 Pandemic in Recife,

Northeastern Brazil. *J Aerosol Science and Engineering, Springer* 6, 316-322 (2022).

30. Junji, C., Yecheng, Z., Quanjiao, C., Maosheng, Y., Rongjuan, P., Yun, W., Yang, Y., Yu, H., Jing, W., Wuxiang, G., 2021. Ozone Gas Inhibits SARS-CoV-2 Transmission and Provides Possible Control Measures. *J Aerosol Science and Engineering, Springer* 5, 516-523 (2021).
31. Talib, D., Dimitris, D., 2020. On respiratory droplets and face masks, *J Physics of Fluids* 32, 063303 (2020).
32. Prather, A. K., Wang, C. C., Schoole, R. T., 2020. Reducing transmission of SARS-CoV-2 Masks and testing are necessary to combat asymptomatic spread in aerosols and droplets, *Science*, 368, 6498 (2020).

Printed by Books on Demand GmbH, Norderstedt / Germany